At Europe's Service

Thomas Jansen • Steven Van Hecke

At Europe's Service

The Origins and Evolution
of the European People's Party

 Springer

Thomas Jansen
thomas.jansen@libero.it

Dr. Steven Van Hecke
University of Antwerp
Sint-Jacobstraat 2
2000 Antwerp
Belgium
steven.vanhecke@ua.ac.be

Centre for European Studies, Rue du Commerce 20, B-1000 Brussels

The Centre for European Studies (CES) is the official think-tank of the European People's Party dedicated to the promotion of Christian Democrat, Conservative and like-minded political values. For more information please visit the CES website: www.thinkingeurope.eu.

This publication receives funding from the European Parliament.

The European Parliament assumes no responsibility for facts or opinions expressed in this publication or their subsequent use.

ISBN 978-3-642-19413-9   e-ISBN 978-3-642-19414-6
DOI 10.1007/978-3-642-19414-6
Springer Heidelberg Dordrecht London New York

Library of Congress Control Number: 2011927265

Cover design: WMXDesign GmbH

Photograph on the cover: European People's Party 2009

Printed on acid-free paper

Springer is part of Springer Science+Business Media (www.springer.com)

Preface

This year the European People's Party (EPP) celebrates its thirty-fifth anniversary – an excellent moment to look back on what has been achieved thus far. From its birth on 8 July 1976 as the Federation of Christian Democratic parties of the European Communities, the EPP has been a factor in shaping European integration. At the time of its origins, it was one of three partly overlapping party organisations of Europe's centre right, along with the older, continental European Union of Christian Democrats (EUCD) and the European Democratic Union (EDU), a loose association of right-of-centre parties. It was by no means clear in 1976 that by the beginning of the second decade of the twenty-first century the EPP would have grown far beyond its original Christian Democratic core, absorbing the other two organisations and continuing as the strongest political force, both in the European Parliament (EP) and in most of the European Union Member States. Consequently, it has not only become the definitive party family on the centre-right, but also a truly European party, closely cooperating with, but distinct from, the EPP Group in the EP. This happened earlier in the EPP's development than in that of the respective organisations of the Liberals, Socialists and Greens.

Today, the EPP family consists of Christian Democrats and many other non-collectivist strands, encompassing a broad range of political tendencies and identities of the centre-right. Many of them, both in Western Europe and especially in Central and Eastern Europe, do not fit the classic labels of Christian Democrats, Conservatives and Liberals, but are woven of combinations of those threads. In that sense the EPP in its own expansion mirrors the enlargement of the EU, and the strengthening of EPP structures over time mirrors the institutional deepening of integration over recent decades. The EPP has always been, more than anything else, essentially at Europe's service.

The EPP family has faced many decisive challenges. From the fall of the Berlin Wall, the collapse of Communism and the unification of Germany, to the accession of Central Europe's new democracies to NATO and the EU, to the financial and economic crisis, the EPP has time and again played a key role in defusing conflicts of interest, creating synergies between people and between institutions, always taking European integration another decisive step forward. This skill is again in demand in the severe crisis that surrounds the euro today. We can prevail only by relying on the fundamental values that have nurtured us through our history: the Christian image of man as a point of departure, human dignity, freedom in responsibility, solidarity, subsidiarity – and the recognition that only a strong EU guarantees Europe's future in the world. Drawing on these values, the EPP family will be able to master present and future challenges: the euro crisis, climate action, demographic change, immigration and integration, and regaining competitiveness in the globalised economy.

This book describes how the EPP was created and how it developed into its present shape and structure. Co-authored by Thomas Jansen, the former Secretary General of the EPP, and Steven van Hecke, Senior Research Fellow at the University of Antwerp, the book is based on an original text from 1996 by Jansen, which appeared in an updated version in 2006. For the 2011 edition, the entire text has been reviewed, restructured and once more updated, and now includes an account of the creation of the EPP's political foundation, the Centre for European Studies (CES), in 2007.

I have been closely involved with the development of the EPP as described in this book, not only as its President since 1990 but also in its founding stage in 1975 and 1976. I am proud to present this revised text on the origins and evolution of the EPP. My thanks go to the authors for their tireless work, and to the CES for editing and coordinating this volume.

Wilfried Martens, EPP President and CES President,
Brussels, March 2011

Acknowledgements

This book is a revised and updated version of *The European People's Party: Origins and Development* by Thomas Jansen (2006), published by the EPP in English, French, German, Italian and Spanish on the occasion of the party's thirtieth anniversary. The 2006 edition was based on the original version of the book, published in German by European Union Verlag in 1996, then translated into English by Edward Steen and published by Macmillan in 1998.

The current revising and updating (to 1 March 2011) was done by Steven Van Hecke, assisted by Tom Roels. For this purpose interviews were conducted with EPP President Wilfried Martens (1 September 2009), Secretary General Antonio López-Istúriz (13 January 2010), Deputy Secretary General Christian Kremer (17 December 2009) and Deputy Secretary General Luc Vandeputte (13 January and 10 February 2010). Chapter 12, on the establishment of the Centre for European Studies (CES), is a new chapter and was written by Tomi Huhtanen, with research support from Elaine Larsen. The text of the book was approved by Thomas Jansen and edited by Marvin DuBois and the Communicative English editing team. Nicholas Alexandris, Roland Freudenstein, Katarína Králiková, Vít Novotný and other CES staff members were involved in the revision of the original version of the book. The final responsibility for the book lies with the CES.

Authors

Thomas Jansen was Secretary General of the European People's Party (EPP) and the European Union of Christian Democrats (EUCD) between 1983 and 1994. Before he took up this position, he had been head of the Rome office of the Konrad Adenauer Foundation. After serving as Secretary General, Jansen worked in the Forward Studies Unit of the European Commission. Before his retirement in 2004, he served as head of the Office of the President of the European Economic and Social Committee (EESC).

Steven Van Hecke is a Senior Research Fellow at the Political Science Department of the University of Antwerp and at the KADOC Centre for Religion, Culture and Society of the Katholieke Universiteit Leuven. He teaches comparative and European Union politics. His area of research is European integration and political parties, and he has also published on transnational party federations.

Contents

Abbreviations and Acronyms

ACDP	Archiv für Christlich-Demokratische Politik
AKP	Adalet ve Kalkınma Partisi (Turkey)
ALDE	Alliance of Liberals and Democrats for Europe
AN	Alleanza Nazionale (Italy)
AP	Alianza Popular (Spain)
ARP	Anti-Revolutionaire Partij (the Netherlands)
AWS	Akcja Wyborcza Solidarność (Poland)
BPF	Bielaruski narodny front (Belarus)
CCD	Centro Cristiano Democratico (Italy)
CDA	Christen Democratisch Appèl (the Netherlands)
cdH	centre démocrate Humaniste (Belgium: French-speaking)
CDI	Christian Democrat International
CDICID	Christian Democrat International Centre for Information and Documentation
CDS	Centre des démocrates sociaux (France)
CDSp[1]	Centro Democrático e Social (Portugal)
CDS-PP	Centro Democrático e Social–Partido Popular (Portugal)
CDU	Christlich Demokratische Union (Germany)
CDUCE	Christian Democratic Union of Central Europe
CDUi[2]	Cristiani Democratici Uniti (Italy)
CD&V	Christen-Democratisch & Vlaams (Belgium: Dutch-speaking)
CDWU	Christian Democratic World Union
CEPESS	Centrum voor Economische, Politieke en Sociale Studies

---

[1] The official abbreviation for 'Centro Democrático e Social' is 'CDS'. This book uses the abbreviation 'CDSp' to avoid confusion with the Centre des démocrates sociaux.

[2] The official abbreviation for 'Cristiani Democratici Uniti' is 'CDU'. This book uses the abbreviation 'CDUi' to avoid confusion with the Christlich Demokratische Union.

CES	Centre for European Studies
CET	Centre for Political Parliamentary Education and Training
CHU	Christelijk-Historische Unie (the Netherlands)
CoE	Council of Europe
CoR	Committee of the Regions
ČSL	Československá strana lidová (Czechoslovakia)
CSU	Christlich-Soziale Union (Germany: Bavaria)
CSV	Chrëschtlech-Sozial Vollekspartei (Luxembourg)
CVP	Christelijke Volkspartij (Belgium: Dutch-speaking)
CVP/PDC	Christlich Demokratische Volkspartei der Schweiz/Parti démocrate-chrétien suisse (Switzerland)
DC	Democrazia Cristiana (Italy)
DCc[3]	Demokratski centar (Croatia)
DEMYC	Democratic Youth Council
DISY	Dimokratikos Synagermos (Cyprus)
DP	Demokratičeska Partija (Bulgaria)
DSB	Demokrati za Silna Bălgarija (Bulgaria)
DSS	Demokratska stranka Srbije (Serbia)
EAEC	European Atomic Energy Community
EALRER	European Association of Locally and Regionally Elected Representatives
EC	European Community
ECR	European Conservatives and Reformists
ECSC	European Coal and Steel Community
EDA	European Democratic Alliance
EDC	European Defence Community
EDG	European Democratic Group
EDS	European Democrat Students
EDU	European Democratic Union
EESC	European Economic and Social Committee
EFLP	European Federation of Local Politicians
ELDR	European Liberal Democrat and Reform Party
ELRPU	European Local and Regional Political Union
EP	European Parliament
EPC	European Political Cooperation
EPP	European People's Party
EPP-ED Group	Group of the European People's Party (Christian Democrats) and European Democrats
ESCU	European Senior Citizens' Union

---

[3] The official abbreviation for 'Demokratski centar' is 'DC'. This book uses the abbreviation 'DCc' to avoid confusion with the Democrazia Christiana.

EU	European Union
EUCD	European Union of Christian Democrats
EUCDW	European Union for Christian Democratic Workers
EUYCD	European Union of Young Christian Democrats
EYCD	European Young Christian Democrats
FAES	Fundación para el Análisis y los Estudios Sociales
FG	Fine Gael (Ireland)
FI	Forza Italia (Italy)
FKgP	Független Kisgazdapárt (Hungary)
FPÖ	Freiheitliche Partei Österreichs (Austria)
GERB	Grazhdani za Evropeysko Razvitie na Balgariya (Bulgaria)
HDZ	Hrvatska demokratska zajednica (Croatia)
HDZBiH	Hrvatska demokratska zajednica Bosne i Hercegovine (Bosnia and Herzegovina)
HSS	Hrvatska seljačka stranka (Croatia)
ICDU	International Christian Democratic Union
IDU	International Democratic Union
IGC	Intergovernmental Conference
IRL	Isamaa ja Res Publica Liit (Estonia)
IRI	International Republican Institute
JL	Jaunais Laiks (Latvia)
JU	Junge Union (Germany)
KAS	Konrad-Adenauer-Stiftung
KDH	Kresťanskodemokratické hnutie (Slovakia)
KDNP	Kereszténydemokrata Néppárt (Hungary)
KD	Kristdemokraterna (Sweden)
KDS	Kristdemokratiska Samhällspartiet (Sweden)
KDScz[4]	Křesťanskodemokratická strana (Czech Republic)
KDU-ČSL	Křesťanská a demokratická unie–Československá strana lidová (Czech Republic)
KOK	Kansallinen Kokoomus (Finland)
KVP	Katholieke Volkspartij (the Netherlands)
LKD	Lietuvos krikščionys demokratai (Lithuania)
MDF	Magyar Demokrata Fórum (Hungary)
MEP	Member of the European Parliament
MOC	Mouvement ouvrier chrétien (Belgium: French-speaking)
MRP	Mouvement républicain populaire (France)
MS	Moderata Samlingspartiet or Moderaterna (Sweden)

---

[4] The official abbreviation for 'Křesťanskodemokratická strana' was 'KDS'. This book uses the abbreviation 'KDScz' to avoid confusion with the Kristdemokratiska Samhällspartiet.

NATO	North Atlantic Treaty Organisation
ND	Néa Dēmokratía (Greece)
NEI	Nouvelles équipes internationales
NSi-KLS	Nova Slovenija–Krščanska ljudska stranka (Slovenia)
NSNU	Narodnyi Soyuz Nasha Ukrayina (Ukraine)
ODA	Občanská demokratická aliance (Czech Republic)
ODCA	Organización Demócrata Cristiana de América
ODS	Občanská demokratická strana (Czech Republic)
OSCE	Organisation for Security and Cooperation in Europe
ÖVP	Österreichische Volkspartei (Austria)
PACE	Parliamentary Assembly of the Council of Europe
PD	Partidul Democrat (Romania)
PDa	Partia Demokratike e Shqipërisë (Albania)
PDCS	Partito Democratico Cristiano Sammarinese (San Marino)
PdL	Il Popolo della Libertà (Italy)
PD-L	Partidul Democrat–Liberal (Romania)
PDP	Partido Demócrata Popular (Spain)
Pdp[5]	Parti démocrate populaire (France)
PDPbih[6]	Partija demokratskog progresa (Bosnia and Herzegovina)
PES	Party of European Socialists
PiS	Prawo i Sprawiedliwość (Poland)
PLDM	Partidul Liberal Democrat din Moldova (Moldova)
PN	Partit Nazzjonalista (Malta)
PNŢCD	Partidul Naţional Ţărănesc Creştin Democrat (Romania)
PNV	Partido Nacionalista Vasco (Spain: the Basque Country)
PO	Platforma Obywatelska (Poland)
PP	Partido Popular (Spain)
PPCD	Partidul Popular Creştin Democrat (Moldova)
PPI	Partito Popolare Italiana (Italy)
PS	Pilsoniskā Savienība (Latvia)
PSC	Parti social chrétien (Belgium: French-speaking)
PSChD	Polskie Stronnictwo Chrześcijańskiej Demokracji (Poland)
PSD	Partido Social Democrata (Portugal)
PSL	Polskie Stronnictwo Ludowe (Poland)
RI	Rinnovamento Italiano (Italy)

---

[5] The official abbreviation for 'Parti démocrate populaire' is 'PDP'. This book uses the abbreviation 'Pdp' to avoid confusion with the Partido Demócrata Popular.

[6] The official abbreviation for 'Partija demokratskog progresa' is 'PDP'. This book uses the abbreviation 'PDPbih' to avoid confusion with the Partido Demócrata Popular.

RMDSz	Romániai Magyar Demokrata Szövetség (Romania)
RPR	Rassemblement pour la République (France)
RS	Ruch Społeczny (Poland)
RUKH	Narodnyi rukh Ukraïny (Ukraine)
SDA	Stranka demokratske akcije (Bosnia and Herzegovina)
SDKÚ	Slovenská demokratická a kresťanská únia (Slovakia)
SDKÚ-DS	Slovenská demokratická a kresťanská únia–Demokratická strana (Slovakia)
SDS	Slovenska demokratska stranka (Slovenia)
SIPDIC	Secrétariat international des partis démocratiques d'inspiration chrétienne
SKD	Slovenski Krščanski Demokrati (Slovenia)
SKL	Stronnictwo Konserwatywno-Ludowe (Poland)
SLS	Slovenska ljudska stranka (Slovenia)
SME UNION	Small and Medium Entrepreneurs' Union
SMK-MKP	Strana maďarskej koalície–Magyar Koalíció Pártja (Slovakia)
SVP	Südtiroler Volkspartei (Italy: South Tirol)
S&D	Socialists and Democrats for Europe
TOP09	Tradice, Odpovědnost, Prosperita 09 (Czech Republic)
TP	Tautas Partija (Latvia)
TS	Tėvynės Sąjunga (Lithuania)
TS-LKD	Tėvynės Sąjunga–Lietuvos krikščionys demokratai (Lithuania)
UCP	Abjadnanaja hramadzianskaja partyja Bielarusi (Belarus)
UDC	Unió Democràtica de Catalunya (Spain: Catalonia)
UdC	Unione di Centro (Italy)
Udeur	Unione Democratici per l'Europa (Italy)
UDF	Union pour la démocratie française (France)
UDFb[7]	Săjuz na Demokratičnite Sili (Bulgaria)
UDPV	Unión Democrática des Pais Valenciano (Spain: Valencia)
UIJDC	Union internationale des jeunes démocrates-chrétiens
UMP	Union pour un mouvement populaire (France)
UNM	Ertiani Natsionaluri Modzraoba (Georgia)
UW	Unia Wolności (Poland)
VMRO-DPMNE	Vnatrešna Makedonska Revolucionerna Organizacija–Demokratska Partija za Makedonsko Nacionalno Edinstvo (FYROM)

[7] The official English abbreviation for 'Săjuz na Demokratičnite Sili' is 'UDF'. This book uses the abbreviation 'UDFb' to avoid confusion with the Union pour la démocratie française.

VMSZ	Savez vojvođanskih Mađara–Vajdasági Magyar Szövetség (Serbia)
WEU	Western European Union
YEPP	Youth of the European People's Party
ZNS	Zemedelski Naroden Sajuz (Bulgaria)

List of Tables

Part I

Origins and Development

Chapter One

Paving the Way: The SIPDIC, NEI and EUCD

The European People's Party (EPP) did not appear out of the blue in 1976. It developed out of diverse forms of cooperation that had long existed among Christian Democrats in Western Europe. The first institutionalised cooperation dates back to the early twentieth century, when the International Secretariat of Christian-inspired Democratic Parties (Secrétariat international des partis démocratiques d'inspiration chrétienne, SIPDIC) provided opportunities for contact among Christian Democratic politicians from Germany, France, Italy and the Benelux. After the Second World War the SIPDIC was replaced by the New International Teams (Nouvelles équipes internationales, NEI). Representatives were grouped into national *équipes* (teams) that represented one or more parties from a given country. In 1965 the NEI renamed itself the European Union of Christian Democrats (EUCD). As an organisation of national political parties from various European countries, it was the immediate forerunner of the EPP.[8]

The Secrétariat International des Partis Démocratiques d'Inspiration Chrétienne (1925–39)

Political parties of Christian inspiration first made contact with each other in the early 1920s, having begun to establish themselves in the nineteenth century at a time when various European countries were progressively em-

---

[8] The archives of the NEI, of the EUCD and of the EPP are located in the Archive for Christian Democratic Policy of the Konrad Adenauer Foundation (Archiv für Christlich-Demokratische Politik/Konrad-Adenauer-Stiftung, ACDP/KAS) in Sankt Augustin, near Bonn.

bracing democracy. Others joined later, after the First World War. By contrast, the earliest predecessor of the modern Socialist International was instrumental in inspiring and organising many of the world's Socialist parties after its founding. The international Christian Democratic movement, however, was the offspring of the national parties.[9]

The experiences of the First World War, coupled with the looming threat of Fascism in Central and Western Europe in the 1920s, led to the conviction among the leaders of Christian-based parties that overcoming nationalism was the decisive precondition for preserving the peace. All political parties, especially Christian-based ones, had a moral obligation to make a concrete contribution to this goal. The first initiative came from the Italian Catholic priest Luigi Sturzo, who had established the Italian People's Party (Partito Popolare Italiano, PPI) in 1919, in which he had also engaged Alcide de Gasperi. Sturzo's involvement was no coincidence: he had personally experienced Fascism and had gone into exile after Mussolini seized power.

Luigi Sturzo

In the autumn of 1924, while a refugee in London, Sturzo contacted the leaders of Christian parties in different European countries to organise a meeting. A few years before, he and a PPI delegation had visited the offices of various party leaders to discuss the possibility of founding an International of People's Parties (Internazionale Popolare).[10] Sturzo was de-

---

[9]   See, among others, Jean-Marie Mayeur, *Des Partis catholiques à la démocratie chrétienne* (Paris, 1980) and Michael P. Fogarty, *Christian Democracy in Western Europe 1820–1953* (London, 1957). For Germany, Austria, Belgium and the Netherlands, see the relevant chapters in Winfried Becker and Rudolf Morsey (eds.), *Christliche Demokratie in Europa: Grundlagen und Entwicklungen seit dem 19. Jahrhundert* (Cologne and Vienna, 1988).

[10]   For Sturzo's contacts and his efforts to spread the idea of the International of People's Parties, and also in particular on the founding and activities of the SIPDIC, see the description by Alwin Hanschmidt, with its numerous sources and references, in 'Eine christlich-demokratische "Internationale" zwischen den Weltkriegen: Das "Secrétariat International des Partis Démocratiques d'Inspiration Chrétienne" in Paris', in Becker and Morsey, *Christliche Demokratie,* 158–88. See also Roberto Papini, *L'Internationale démocrate-chrétienne: la coopération entre les partis démocrates-chrétiens de 1925 à 1986* (Paris, 1988), 31ff. Roberto Papini's 'Il coraggio della democrazia: Luigi Sturzo e l'Internazionale Popolare tra le due guerre' (unpublished paper,

termined to establish international cooperation among all Christian-based parties. During the 1920s, however, any party that could realistically be considered a possible partner or was prepared to cooperate in the way Sturzo had in mind was, without exception, Catholic. These were either parties made up of Catholic members or parties that represented political Catholicism. Catholicism's inherent internationalism – the international character of the church and its natural ties across borders – encouraged such party leaders to cooperate with each other.

The meeting initiated by Sturzo took place in Paris on 12–13 December 1925. The invitation came from the Popular Democratic Party (Parti démocrate populaire, Pdp), which had been founded only the previous autumn. Along with the French Pdp, the Italian PPI, the Belgian Christian Workers' League (Mouvement ouvrier chrétien, MOC), the Polish Christian Democratic Party (Polskie Stronnictwo Chrześcijańskiej Demokracji, PSChD) and the German Centre Party (Deutsche Zentrumspartei) were represented. The delegates from the five countries were all senior figures. Invitations to like-minded parties in the Netherlands, Switzerland, Czechoslovakia, Lithuania and Spain were not taken up.

The delegates who came to Paris reported on their parties in general, their principles and their political tasks. With 'due caution and restraint', a debate took place on the political situation and developments in Europe, the twin menaces of Fascism and Communism, and current foreign policy issues.[11] It soon became clear how difficult it was, given national allegiances, to agree on common positions. Nor was a single one reached. Indeed, it was agreed that this was impossible, and even the fact that the meeting had taken place was to be kept secret. The French in particular feared that there might be political repercussions inside the Pdp. Such reticence was the opposite of Sturzo's intentions: he had hoped for common action that would have a public *éclat*. As the father of the emerging Christian Democrat International, Sturzo must have understood from this preliminary conference how difficult it would be to move parties beyond simple information exchanges to joint political meetings or joint action.[12]

1995) is particularly helpful, since it provides a good deal of information on the whole period dealt with in this chapter, especially Sturzo's philosophy and the development of cooperation among Christian Democrats in Europe.

[11] Raymond Laurent, Secretary General of the Pdp, quoted by Hanschmidt, 'Eine christlich-demokratische "Internationale"', 168.

[12] Ibid. 169ff.

The discussion in 1925 highlighted two opposing concepts of future co-operation: the Italian idea of a permanent interparliamentary union and the French proposal of a *bureau central d'information*. Eventually the French proposal was accepted and it was decided that an international secretariat of parties (or political organisations) inspired by Christianity should be established in Paris. Its task would be to organise a network and a flow of information between parties. Another meeting was to take place in Brussels in May 1926, which parties not at the Paris meeting would be encouraged to attend.

The Brussels meeting confirmed and formalised the substance of what had been decided in Paris, although on this occasion only four of the five founding parties were represented; the Polish delegation was unable to attend. There were also apologies for the absences of the Lithuanian Christian Democratic Party (Lietuvos krikščionys demokratai, LKD), the Austrian Christian Social Party (Christlichsoziale Partei), the Swiss Catholic Conservative Party (Katholisch-Konservative Partei der Schweiz) and the Czech People's Party (Československá strana lidová, ČSL), which were counted as applications for membership, since these parties more or less regularly took part in subsequent meetings.[13]

The French Influence

The Paris secretariat was now officially named the SIPDIC. The French conception of a *bureau* had triumphed over the Italian proposal of a unified body. Although the bodies responsible for the work of the SIPDIC were to be made up of officials from the parties, not individual figures, according to the statutes, 'this grouping is not, at the present time, aimed at creating a federation of the participating parties.'[14]

Even the organisation's name, which avoided the term 'Christian Democratic', can be traced to French influence, which would remain decisive. Since Paris had been chosen as the seat of the SIPDIC, it also became the

---

[13] Parties from Luxembourg, Hungary and the Netherlands (Roomsch-Katholieke Staatspartij) joined later. See the list of SIPDIC member parties in Hanschmidt, 'Eine christlich-demokratische "Internationale"', 187ff., which also indicates that apart from the parties mentioned above there were relationships with kindred political figures and groups in Spain, Yugoslavia and Romania.

[14] Hanschmidt, 'Eine christlich-demokratische "Internationale"', 172: 'le groupement ne vise pas actuellement à constituer une Fédération internationale des partis représentés' [translation from 1998 version of the book].

Pdp's job to staff it. To begin, Raymond Laurent and Philippe de Las Cases led the SIPDIC, followed by Henri Simondet from 1928 until the SIPDIC folded.[15]

A *comité central* formed the political executive body; later its name was changed to *comité exécutif*. Each party had a delegate to this committee, with the right to send a substitute. The executive committee usually met twice a year, one of the meetings taking place alongside a conference or party congress of one of the SIPDIC parties in the years between 1926 and 1932. Member parties would each send several delegates to these larger meetings.[16]

All such meetings and sessions were private. The decisions or conclusions reached over the years on topical or fundamental issues were confidential, intended only for the internal use of the member parties. There was a single exception to this rule. In January 1931, the executive committee published a declaration on securing peace that referred to 'the difficult international situation caused by rigid nationalism and the difficult economic crisis'. The statement condemned any use of force, in support of either domestic or foreign policy, and affirmed member parties' commitment to an 'effective organisation of peace by the League of Nations, and to a rapprochement between peoples which must be achieved through both economic and political cooperation'.[17]

Simondet justified the SIPDIC's departure from its principle and previous practice by saying it was necessary to make a public declaration or the organisation would be untrue to its task. Its efforts would appear fruitless if member parties did not, in the current circumstances, publicly show themselves to be 'une force au service de l'oeuvre de paix et de collaboration internationale'.[18] Another reason for the SIPDIC to make a public statement was to prevent public opinion from becoming used to the idea that the movement for peace and rapprochement between peoples was something that concerned only socialists, radicals and the extreme left.

---

[15] On the role, orientation and contribution of the Pdp, see Jean Claude Delbreil, 'Les Démocrates d'inspiration chrétienne et les problèmes européens dans l'entre-deux-guerres', in Serge Berstein, Jean-Marie Mayeur and Pierre Milza, *Le MRP et la construction européenne* (Brussels, 1993), 15–39.

[16] For a list of SIPDIC conferences and congresses and information about the participating delegations, see Hanschmidt, 'Eine christlich-demokratische "Internationale"', 186ff.; see also Roberto Papini's account in *L'Internationale*, 35ff.

[17] Hanschmidt, 'Eine christlich-demokratische "Internationale"', 179.

[18] Ibid. 180.

The International Context

Remonstrations and proposals from Italian representatives were instrumental in ensuring that there was more and more discussion about the SIPDIC's mission; however, no change was made to the SIPDIC's limited role as an information bureau only. Initiatives aimed at stronger organisational cooperation were not pursued. One reason, no doubt, was to avoid giving any impression of imitating the example of the Socialist International. In general, especially among parties from smaller countries, there was a degree of nervousness and diplomatic caution about any statement that could be interpreted as interference in the politics of a foreign power. In these circumstances it was scarcely possible to develop policies beyond general statements of principles.

The SIPDIC's member parties were also pursuing different agendas, and there was no procedure for producing consensus: 'While the Italians wanted to give its structure an unambiguously anti-fascist character, and the French above all sought dialogue with the Germans, most feared more than anything else having to adopt too clear a political position; they preferred to deal with the traditional, less political, themes of Christian social politics (the family, workers' profit-sharing schemes, etc.).'[19]

The time had not yet come for creating international democratic organisations that could act together to defend common political positions. Even at the national level, democracy and its institutions were extremely weak. In Italy they had already been crushed; in Austria and Germany they would not survive much longer.

A hardened nationalism dominated inter-war Europe and left practically no space for internationalism. International relations and 'transnationalism' could survive and be practised to an extent, but only discreetly, as through they were a shameful activity. Proponents of international co-operation were widely suspected of being unpatriotic. It was with discretion – privately and confidentially – that the 'International' of Catholic Christian Democratic parties tried to operate.[20]

Against such a backdrop, it took a special far-sightedness – something few politicians had at the time – to realise that democracy, democratic forms and democratic procedures also made sense as a basis for political relations in the international context. Without question, Sturzo was one of

---

[19]   Papini, *L'Internationale*, 36 [translation from 1998 version of the book].
[20]   Hanschmidt, 'Eine christlich-demokratische "Internationale"', 184.

those who possessed it. But his proposals and representations mostly fell on deaf ears. He complained to Francesco Luigi Ferrari, who in 1929 succeeded him as the representative of the Italian PPI, that 'the endeavours made from the beginning to give [the SIPDIC] an international democratic orientation have progressively lost their influence.'[21] Transnational relations between parties therefore remained stuck in the tramlines of international diplomacy. The League of Nations, which during the 1920s and 1930s was the institutional framework for international cooperation between states, offered democratic forces no way to make their influence felt. It was not a league of nations but a diplomatic organisation.

The European Perspective

The most active and serious of those taking part in the work of the SIPDIC were the representatives from member parties in Belgium, France, Italy, Luxembourg, the Netherlands and (until 1932) Germany. It is hardly a coincidence that these were the countries that would later found the European Community (EC). After the Second World War, Christian Democratic politicians from these countries who began the process of European integration could draw strength from the common lessons and experiences of cooperation between 1925 and 1939.

The evidence of congresses organised by the SIPDIC shows, at this much earlier stage, both a well-developed sensibility to and positive interest in uniting Europe and beginning to work towards integration. In the 1930s, all this was politically impossible. But the SIPDIC's last Congress, held in Cologne in October 1932 – a few months before the National Socialists seized power in Germany – under the chairmanship of Konrad Adenauer, did vote in favour of this final declaration:

We must strengthen and encourage comprehensive cooperation between all European nations in order to achieve a Common Market for production and the free movement and consumption of goods . . . Full union, which is the final goal, cannot be achieved immediately or directly. So we must gradually remove customs barriers, and trade and financial barriers preventing the regular exchange of

---

[21] Quoted by Papini in *L'Internationale*, 36 [translation from 1998 version of the book].

goods, in order to realise as quickly as possible the free movement of goods, capital, and people.[22]

This proposal for overcoming the crisis in Europe was the forerunner of those projects that after the Second World War led to the creation of the EC.

A Temporary Halt

After 1933 it was impossible for the SIPDIC to organise any more large-scale conferences. There were still fairly regular meetings of a more limited kind, mostly in Paris, but they had no effect in terms of politically relevant action. Neither the tools nor the political will existed to advance the movement. People stayed in contact and kept each other informed. An initiative to re-establish the SIPDIC, again coming from Sturzo, found no support. The SIPDIC's last sign of life was a confidential memorandum to Pope Pius XI in January 1938, asking him to speak out about the dangers facing Europe and to use his moral authority to rescue the peace.

Fascism, National Socialism, increased tensions between governments, the spirit of revenge and the dictator's obsession with power all eventually led to the Second World War, ending for the moment cooperation between the Christian Democratic parties. The SIPDIC ceased to exist in 1939. The outbreak of war had crushed for the time being any worthwhile democratic efforts to shape the Continent.

In London, in the winter of 1940–41, the International Christian Democratic Union (ICDU) was founded by British and exiled politicians. Once again Sturzo was involved, and an attempt was made to continue the tradition of the SIPDIC. This initiative was linked to the People and Freedom Group, an organisation that had been set up by Sturzo in 1936. Thanks to its personal contacts and disseminating of information, the ICDU made a substantial contribution to maintaining connections among Christian Democratic figures during the war, ensuring they could flourish afterwards. But it could do little more of genuine political relevance. Only after the terrible experience of national totalitarianism, with its contempt for human

---

[22] Cited in Fraktion der Europäischen Volkspartei, *Zur Geschichte der christlich-demokratischen Bewegung in Europa* (Melle, 1990), 128ff. The proposal to create a Common European Market was raised in 1923 – and variants were repeated – by Konrad Adenauer, Oberbürgermeister (Mayor) of Cologne from 1917 to 1933 and President of the Prussian State Council from 1920 to 1933.

beings and its destructiveness, would it be possible to take further steps on the road to international democracy, and with it, international party work.

The Nouvelles Équipes Internationales (1948–65)

Leaders of re-established or newly founded parties with a Christian Democratic orientation made contact with each other soon after the Second World War. Many were old friends from the 1920s and 1930s and had worked together under the arrangements made by the SIPDIC in the period 1925–39. Some had kept up connections in exile or in resistance movements.

The initiative came from the Swiss Catholic Conservative Party (Katholisch-Konservative Partei der Schweiz). Thanks to Swiss neutrality, it had been the only Christian Democrat party to survive the war unscathed. An invitation to a meeting in Lucerne from 27 February to 2 March 1947 was accepted by all those invited. Delegates arrived from Belgium, Great Britain, France, Italy, the Netherlands, Luxembourg, Austria and Switzerland. No one was able to come from Czechoslovakia or Hungary. And no invitation was sent to Germany, where the political situation remained unclear.

There was general agreement at this meeting about both the logic of cooperation and its intrinsic value. But once again there were quarrels, familiar from the founding of the SIPDIC in 1925, about what form such cooperation should take. This time it was the Swiss, supported by the Italians and the Austrians, who argued for an *entente organique* of parties going in the same political direction. By this they meant an organisation capable of hammering out a common position and securing common action among its member parties. The Belgians, supported by the French and Dutch, insisted that there were both domestic and international reasons for limiting an international association to one in which, at most, representative Christian Democratic figures could work together. The parties were not to be involved. This argument finally won the day.[23]

---

[23] See Roberto Papini, 'Les débuts des Nouvelles équipes internationales', in Hugues Portelli and Thomas Jansen (eds.), *La Démocratie chrétienne: force internationale* (Nanterre, 1986), 31–40. See also Papini, *L'Internationale*, 47ff. and Philippe Chenaux, *Une Europe vaticane? Entre le plan Marshall et les traités de Rome*, (Brussels, 1990), 119–56. On the motives of the French and

Naming the organisation also proved controversial. Eventually, follow-ing the determined wishes of the French and Belgians, it was decided that the association should be called the NEI. The Belgians, anticipating this, had already founded an association with this name. They now offered their organisation as a framework and invited everyone to join. Their far-sightedness was rewarded; they were given the task of coordinating the first Congress.

A Curious Name

The curious name deliberately echoed that of the Nouvelles équipes fran-çaises, which just before the war (1938–39) had tried to bring together committed democrats from various French parties and groups.[24] The name also exposed a dilemma, one which would long haunt Christian Demo-cracy's organisation and identity in Europe. At the same time, it pointed to a possible direction that was to prove important later.

The dilemma was that some parties that were prepared to cooperate in the NEI and actually wanted to be Christian Democrats did not wish to be identified by that name. In France, particularly, the term Christian Democ-rat was used only with the greatest reticence, and hardly ever in public. The fear was that it would be misunderstood as referring to something clerical. It was absolutely no part of the Christian Democrats' post-war idea of themselves to be the political wing of a religious authority, nor un-critically to defend the church(es) or their interests. They did not even want to be suspected of this. However, in France at least, no one dared to openly refute the public misunderstanding, created by liberal and socialist propaganda, that Christian Democracy was some kind of right-wing 'Black International'.[25]

The possibility foreshadowed by the name Nouvelles équipes was that a Christian-inspired political movement could be open, and stay open, to

the related controversy in the MRP, see Jean-Claude Delbreil, 'Le MRP et la construction européenne: résultats, interprétations, et conclusions d'une en-quête écrite et orale', in Berstein, Mayeur and Milza, Le MRP, 309–63.

[24] See Robert Bichet, La Démocratie chrétienne en France: Le Mouvement ré-publicain populaire (Besançon, 1980), 27ff.

[25] Emiel Lamberts, The Black International 1870–1878: The Holy See and Mili-tant Catholicism in Europe (Leuven, 2002).

political forces that were like-minded but coming from other ideological traditions. Thus, neutral names were chosen: in France, the post-war Christian Democrat–oriented parties were called the Republican People's Movement (Mouvement républicain populaire, MRP), the Democratic Centre (Centre démocrate) and the Democratic and Social Centre (Centre des démocrates sociaux, CDS).

So what were the *équipes*? They could, for instance, be teams of members and activists formed from the same parties in various countries. They might be other groups, independent of parties, who had come together to work with like-minded people at an international level. In fact, the NEI was not really a union of parties, but a mixture. Parties belonged to it, but alongside them were political figures who had joined national teams. Among them was a series of members of the French MRP and Belgian Christian People's Party (Christelijke Volkspartij/Parti social chrétien, CVP/PSC) who joined as individuals after their parties refused to become corporate members.[26] Robert Schuman, Georges Bidault, Henri Teitgen and André Colin from France, and Paul van Zeeland, August-Edmond De Schryver and Theo Lefèvre from Belgium were all energetic proponents of the NEI. Great Britain, too, was represented by an *équipe* composed of figures from both its main parties, Conservative and Labour.[27] In the Netherlands, the three Christian (confessional) parties – the Catholic People's Party (Katholieke Volkspartij, KVP), the Anti-Revolutionary Party (Anti-Revolutionaire Partij, ARP) and the Christian Historical Union (Christelijk-Historische Unie, CHU) – formed a joint *équipe* that became a

---

[26]   Only in 1960, following the resignation of August-Edmond De Schryver, who had been NEI President for a decade, did the CVP/PSC (then still the joint Flemish and French-speaking Christian Democratic party in Belgium) become a member of the NEI. On the role of the CVP/PSC and De Schryver, see Philippe Chenaux, 'La contribution belge à la démocratie chrétienne internationale' (unpublished paper, 1995).

[27]   See Karl Josef Hahn and Friedrich Fugmann, 'Die Europäische Christlich-Demokratische Union zwischen europäischen Anspruch und nationalen Realitäten', in W. Wessels, *Zusammenarbeit der Parteien in Westeuropa. Auf dem Weg zu einer neuen politischen Infrastruktur?* Institute for European Politics (Bonn, 1976), 255ff. This article also indicates that in the 1960s Labour members withdrew from the British *équipe* because their party no longer tolerated individual party members holding dual international membership. The effect of this was to pull the rug out from under the supranational British *équipe* and so, in effect, to undermine all political delegations from Great Britain.

member. The constitution provided for only one *équipe* per country. 'Like-minded national groups forced to work in exile' also counted as *équipes*.[28]

The founding Congress of the NEI took place in Chaudfontaine, near Liège, between 31 May and 3 June 1947. As in Lucerne, not only were Western European parties present but also the Polish Labour Union (Stronnictwo Pracy), already in exile, and the ČSL from Czechoslovakia. The Frenchman Robert Bichet was elected President, and the Belgian Jules Soyeur, Secretary General.[29] The goal of the association was laid down in the constitution:

[T]o arrange regular contacts between political groups and personalities from various countries who are informed by the principles of Christian Democracy; they will scrutinize both their own national situations and international problems in the light of these principles, compare experiences and political programmes and, following international agreement, strive to achieve democracy and social and political peace.[30]

The leadership consisted of the executive committee (*comité directeur*), which met three times a year and was composed of representatives of each national *équipe*. Decisions were reached by a two-thirds majority of those present, but there was a determined attempt to achieve consensus. The ruling committee elected a presidium (*bureau politique*) annually. This consisted of the president, four vice-presidents and the secretary general. To begin with, the NEI Secretariat was based in Brussels, and after 1950, in Paris.

Besides those of the executive committee and the presidium, there were intermittent meetings of a cultural committee, an economic and social committee, and an East–West committee. Working groups concerned themselves with parliamentary matters, propaganda, international political problems and political programmes. The International Union of Young Christian Democrats (Union internationale des jeunes démocrates-chrétiens, UIJDC) was established in 1951 as the successor to the NEI's

---

[28] The text of the statute is cited in *Zur Geschichte*, 121ff. Exile groups from Poland, Hungary, Czechoslovakia, the Basque country, Romania and Bulgaria were NEI members.

[29] Bichet's successors were also CVP/PSC members: De Schryver (1950–59) and Theo Lefèvre (1960–65); Soyeur's successors, like him, were MRP members: Bichet (1950–54), Alfred Coste-Floret (1954) and Jean Seitlinger (1955–62).

[30] NEI constitution, reprinted in *Zur Geschichte,* 121ff.

youth commission, founded in 1947. Attempts to issue regular publications were frustrated by financial difficulties.

What was new in the NEI compared with the pre-war SIPDIC was first of all the ecumenical element. This made possible both reconciliation and reconstruction out of the ruins of the national states. What was completely new was the vision of the future, the union of Europe, which led to overcoming the old political system. Finally, the post-war Christian Democratic parties were different from their 1920s predecessors in one crucial way: they were genuine people's parties, having emerged through elections as the leading forces in their countries and having taken on government responsibilities.

The Goal: Uniting Europe

The NEI was also a co-founder and an active partner of the International European Movement. The NEI was among the initiators and organisers of the legendary Europe Congress in The Hague in May 1948.[31] The NEI described the task at hand as being of 'especial urgency following the end of the Second World War: on the one hand, the realisation of European union as a first step to a united world; on the other hand, the union of people and parties inspired by Christian Democracy. In this way, with the help of vital international organisations, [the NEI] wants to make an effective contribution to an enduring peace.'[32]

A large number of conferences and colloquia followed the founding Congress in Chaudfontaine. In truth, the NEI's main activity consisted of organising the annual Congresses and study conferences at which topical European issues were discussed. The resonance of these meetings was

---

[31] For further information on this matter, and on the subject of this chapter in general, see Nicole Bacharan-Gressel, 'Les organisations et les associations pro-européennes', in Berstein, Mayeur and Milza, *Le MRP*, 41ff. See also Heribert Gisch, 'Die europäischen Christdemokraten (NEI)', in Wilfried Loth (ed.), *Die Anfänge der europäischen Integration 1945–1950* (Bonn, 1990), 227–36 and Philippe Chenaux, 'Les Nouvelles équipes internationales', in Sergie Pistone (ed.), *I Movimenti per l'unità Europea dal 1945 al 1954: Atti del convegno internazionale Pavia 19–20–21 ottobre 1989*, Fondazione Europea Luciano Bolis (Milano, 1992), 237–52.

[32] Cited in *Zur Geschichte*, 114–20.

considerable in the first years; directly or indirectly they involved members of governments, party leaders and other important figures.[33]

The agendas for these meetings reveal not only the major problems of the time, but also the expectations that were entertained and particularly the hopes invested in a European union that could both bring peace and preserve it. However, the records also demonstrate that cooperation in the NEI in those days was mainly about trying to make progress by means of classical diplomacy. That corresponded with prevailing conditions at the state level. It was only later in the 1950s, and very gradually, that the perspective changed. This was after the European Coal and Steel Community (ECSC) had been successfully established, and people were prepared to build on it. Only then did awareness of the possibility of integration grow, and alongside it, the vista of supranational and federalist inter-party cooperation as well.

These first steps in political cooperation under the aegis of the NEI are marked by the situation at the time the organisation was founded. Following the experience of national totalitarianism and the catastrophe of the Second World War, Christian Democratic parties in many parts of the Continent offered a convincing spiritual and political alternative. This was the reason Christian Democracy became a decisive political movement. As majority or governing parties, they took responsibility for the political and socio-economic reconstruction of their countries. They were increasingly determined to pursue the political integration of European national states into a supranational community and they sought common security in the Atlantic Alliance.

Agreement on such key questions had a powerful, cohesive effect: developing a clear common political line became possible, despite the absence of an organisational infrastructure. Stabilisation and progress were given a significant boost by joint political activity aimed at the common goal of uniting Europe during an extremely difficult time. Europe had been shattered by war and suffered in the 1940s and 1950s from raw nerves and numerous insecurities. International cooperation between Christian Democratic parties was gaining coherence during the first post-war years, and

---

[33] Documentation of the Congresses and decisions of international Christian Democratic organisations can be found in *La Démocratie chrétienne dans le monde: résolutions et déclarations des organisations internationales démocrates chrétiennes de 1947 à 1973* (Rome, 1973). Resolutions of the 16 NEI Congresses (1947–63) can also be found in *Zur Geschichte*, 156ff.

the NEI was far more cohesive and substantive than the SIPDIC had been between the wars.

In its first years the development of the NEI was accompanied by informal, confidential meetings at the highest political level; these became known as the 'Geneva Circle'.[34] Here, Adenauer and Bidault met for the first time, as did other Christian Democratic politicians who were either in government or were party leaders in their countries. It was at these meetings that NEI cooperation was given a degree of political underpinning. In particular, these meetings prepared the ground politically for the reconciliation of France and Germany. European union was propelled onwards by the success of the understandings reached in this way about the peculiarly delicate questions of the German–French relationship.

Worldwide Cooperation

What is certain is that Christian Democracy's global perspective advanced and influenced cooperation in Europe. The 'European union as a first step towards a united world' was a fundamental driving force for the NEI. Its internationalism was expressed by its name (including the second part, though it was rarely referred to: Union internationale des démocrates chrétiens).[35]

Christian parties were banned in Central and Eastern Europe once Communist rule was imposed. In June 1950 representatives who had fled

---

[34] Ibid. See also Papini, *L'Internationale*, 71ff. and Bruno Dörpinghaus, 'Die Genfer Sitzungen – Erste Zusammenkünfte führender christlich-demokratischen Politiker im Nachkriegseuropa', in Dieter Blumenwitz et al. (eds.), *Konrad Adenauer und seine Zeit: Politik und Persönlichkeit des ersten Bundeskanzlers*. Vol. 1: *Beiträge von Weg- und Zeitgenossen* (Stuttgart, 1976), 358–65; Chenaux, *Une Europe vaticane?*, 128ff.; Michael Gehler, 'Begegnungsort des Kalten Krieges. Der "Genfer Kreis" und die geheimen Absprachen westeuropäischer Christdemokraten (1947–1955)', in Michael Gehler, Wolfram Kaiser and Helmut Worhnout (eds.), *Christdemokratie in Europa im 20. Jahrhundert* (Vienna, 2001), 642–94. The 'Discussions in Geneva' took place between 1947 and 1956, but only until 1952 at the level of leading politicians; once the ECSC had been established, there were numerous opportunities for them to meet.

[35] With regard to questions concerning the name Christian Democrat International, see above all, Papini, *L'Internationale*, as well as pertinent contributions of Hugues Portelli, Brian Palmer, André Louis and Jürgen Hartmann in Portelli and Jansen, *La Démocratie chrétienne*.

to the West, and who had been involved in the NEI from the start, established the Christian Democratic Union of Central Europe (CDUCE). Americans supported the new organisation. It was founded in New York and based its secretariat in Washington D.C. Later on it moved to New York.[36] It included representatives from Czechoslovakia, Poland, Hungary, Lithuania, Latvia and Slovenia/Yugoslavia. Their political, journalistic and propaganda activities were mainly focused on fighting Communism, attacking the Soviet Union and liberating and democratising their countries. Their representatives raised their voices not only in the NEI but also in United Nations bodies, in international Christian associations and organisations and, obviously, in public as well.[37]

Christian Democrats driven out of Central European countries, who had gone into exile in North and South America via Western Europe, as well as political emigrants from Spain under General Franco's dictatorship, were especially important. They contributed considerably to the spread of an intellectual heritage and the establishment of an intercontinental network. They had connections with like-minded groups in the United States, Venezuela, Argentina, Chile and other Latin American countries, as well as with the Christian Democratic Organisation of America (Organización Demócrata Cristiana de América, ODCA), founded in Montevideo in 1947. In the 1950s, those Christian Democrats succeeded in forging the connection between the European and Latin American Christian Democrats.

The first intercontinental conferences involving leading Christian Democrats from Europe and Latin America took place in Paris in 1956 and in Brussels two years later. A third meeting (1961) in Santiago, Chile, saw the foundation of the Christian Democratic World Union (CDWU), which brought together the European NEI, the Central European exile organisation CDUCE, the American regional organisation ODCA and the youth organisation UIJDC.

---

[36]   Roberto Papini, *The Christian Democrat International* (Lanham, 1997), 77.
[37]   See Konrad Siniewicz, 'L'activité internationale des démocrates-chrétiennes de l'Europe centrale', in Portelli and Jansen, *La Démocratie chrétienne*, 233ff. See also Papini, *L'Internationale*, 82ff.

Evolution and Meaning

The German Christian Democrats were already participating by the time of the NEI's second Congress in Luxembourg, held between 28 January and 1 February 1948. The subject was the 'German question'. Adenauer, then scarcely known abroad, led the delegation of the German Christian Democrats (Christlich Demokratische Union, CDU) and spoke about the future shape of Germany in a united Europe. He made a very strong impression, laying the foundation of his future credibility and that of his party and his country. It was the first time that any politician in office in the new Germany had been given a platform at a forum outside his own country. Coming only three years after the end of the war unleashed by the Nazis, this was in no sense routine. There had been internal opposition to German participation in the NEI, though it was quickly overcome. The NEI was the first post-war association to accord a German delegation membership on equal terms.

The Germans became active members of the NEI. As relations developed, they pushed more and more strongly – like the Italians, but against the counterweight of the French – for a stronger organisation and especially for more NEI influence in national parties.[38] The CDU leadership, in particular Adenauer, exploited the possibilities offered by NEI connections as much as possible. This was done both to advance their own interests and to reach agreements on especially contentious issues, notably with the French.

The possible routes along which the NEI could develop were limited by a lack of enthusiasm on the part of the French and Belgians, who shared the job of running it. The organisation, its resources and its possibilities all remained weak. Even Adenauer's impatience could not change that. His analysis, delivered to the CDU leadership in the summer of 1951, was that '[w]e must create a federation with the Christian parties in the other European countries which is able to look after our common interests better than the NEI. The NEI should also have a different name, and be pulled out of

---

[38] See Jean-Dominique Durand, 'Les rapports entre le MPR et la Démocratie chrétienne italienne (1945–1955)', in Berstein, Mayeur and Milza, *Le MRP*, 251ff. See also Reinhard Schreiner, 'La politique européenne de la CDU relative à la France et au MPR des anneés 1945–1966', in ibid. 275ff.

the rut it is in. We should work on creating the Christian International as soon as Ems.'[39]

The NEI was supposed to hold a Congress in Bad Ems on 14–16 September 1951; the CDU federal leadership set up a committee made up of Ministers Jakob Kaiser, Kurt Georg Kiesinger, Wilhelm Simpfendörfer and Adolf Süsterhenn. They were 'to prepare the idea of a Christian International for [the conference at] Ems', but did not manage to do so.[40] Adenauer, who argued forcefully for the International during the Congress, did not get his way here, either:

> But in one respect, I have regretfully to say, we are behind parties which – unlike ourselves – do not believe in the victorious force of Christian thinking. They have been much more effective than we have in organising international co-operation. Please reflect on the Cominform, or on the re-established Socialist International. We, my friends, the Christian parties, cannot at the moment compete with them, either in terms of our strength or as effective propagandists. I openly admit that the German Christian parties' commitment to the common effort has been inadequate. Note, ladies and gentlemen, that the name we go under does not even show what we want. It is a completely neutral name, which can mean absolutely anything. When I think that the Christian parties . . . could forge a stronger alliance by constantly exchanging information and working together: what an amazingly strong effect that would have on what happens in Europe, on the renaissance of Europe . . . Stronger links between the Christian parties would decisively advance our work, work which is in a common cause, and so helps all of us. But above all, stronger cooperation by Christian parties would advance the political integration of Europe.[41]

From the middle of the 1950s onwards, the NEI steadily lost its relevance. With the establishment of the ECSC (1952), and then, crucially, with the foundation of the European Economic Community (EEC, 1958), practical cooperation among Christian Democrats gradually moved to the Christian Democratic groups that had evolved in the Common Assembly,

---

[39] Konrad Adenauer, 'Es musste alles neu gemacht werden', in Günter Buchstab (ed.), *Protocols of the CDU Federal Leadership 1950–1953* (Stuttgart, 1986), 49.

[40] Ibid. 66.

[41] Konrad Adenauer, speech to the NEI, Bad Ems, 14 September 1951, reprinted in Werner Weidenfeld, *Konrad Adenauer und Europa: Die geistigen Grundlagen der Westeuropäischen Integrationspolitik des ersten Bonner Bundeskanzlers*, Studies of the Institute for European Politics, vol. 7 (Bonn, 1976), 326–34; see 331ff.

or European Parliament (EP). Furthermore, several Christian Democratic parties had held on to power throughout this whole period, and their interest in the weak NEI structures had progressively declined. The machinery of government offered such party leaderships perfectly adequate means to communicate and reach agreement with their partners in other countries. The fact that the NEI crisis reached its climax at the beginning of the 1960s is explained by the growing tensions inside the French MRP, and the difficulties created for French Christian Democrats by the return to power of General Charles de Gaulle.[42]

Despite the increasingly obvious weaknesses of its construction, the NEI's efforts during its later years (from 1955 to 1965), contributed a great deal to preserving what the member parties had in common and to the emergence of a consensus about political programmes. The NEI was a vital forum for the identity of Christian Democracy as an influential international force. It created important preconditions for the success of European political integration, and for the development of solidarity between its member parties.

The European Union of Christian Democrats (1965–76)

The EEC, in operation since 1958, proved to be a great success, but attempts to push political integration were met with setbacks. Negotiations over the statute to establish a European Political Union collapsed in 1962, and Italian and German initiatives to breathe new life into these proved fruitless. Moreover, the entry of the United Kingdom into the Community was blocked by de Gaulle.

Despite the tensions that had arisen within the Community, the EEC Commission – energetically led by the German Christian Democrat Walter Hallstein – refused to be flustered and continued with its integration programme. Support came mainly from Christian Democrat–led governments. De Gaulle, who questioned the supranational development of the EC, faced resistance from those who, like the Christian Democrats, remained faithful to the founders' federative model. The break between de Gaulle and the Christian Democrat-oriented French party MRP came in 1962.

---

[42] Delbreil, 'Le MRP', 325.

This was the backdrop to the transformation of the NEI in 1965. It now called itself the 'European Union of Christian Democrats' (EUCD), a name that signalled a change in the character of the organisation. The EUCD was designed to develop into a single, united body whose members pursued common aims. The phase of consultation and cooperation was to be abandoned in favour of 'defining genuine common policies', at the centre of which was 'removing obstacles on the road to creating a European community'.[43] Mariano Rumor, under whose leadership the transformation of the NEI into the EUCD had taken place, observed in hindsight: 'Our cooperation in those days attained a new level of effectiveness and initiative at European level. There also developed intensive and dynamic relations with the parliamentary group; these took on a structured and permanent character. It was because of this development that it was obvious that we had to choose a new, challenging name for our European organisation.'[44]

Renewal and Continuity

With the decline of the French MRP, the secretariat of the NEI, based in Paris since 1950, was moved to Rome in 1964. The relocation took place immediately before the NEI's transformation into the EUCD, and the move itself encouraged bolder thoughts about the form and content of cooperation between Christian Democratic parties. The French element, despite the remarkable contributions of individuals from the ranks of the MRP, had always been strongly focused on national affairs; the organisation had remained Franco-centric. The Italians, by contrast, had from the start argued for a stronger supranational organisation with more authority over member parties.

At a meeting of the executive in Brussels on 3 May 1965, it was decided to make the change to the organisation, and the Congress at Taormina (Italy) in the same year confirmed it. The Belgian deputy Leo Tindemans, elected first Secretary General of the EUCD, drew up the relevant report. For all the intended reforms, the EUCD stressed very clearly not only its continuity but even its common identity with the NEI. The first EUCD Congress, held in Taormina on 9–12 December 1965, was counted as the

---

[43]   Emilio Colombo, 'Internationale Präsenz der Christlichen Demokraten', in *Zur Geschichte*, 80.

[44]   Mariano Rumor, 'Die gemeinsame Aktion der Christlichen Demokraten in Europa', in *Zur Geschichte*, 89.

seventeenth Congress of the NEI; the last NEI Congress on 21–23 June 1962, held in Vienna, was its sixteenth Congress.[45]

The EUCD was given a constitution with stronger unitary elements than those in the NEI's. But as before, member parties retained full autonomy vis-à-vis international bodies, and relations between them were confederal. The members of the organisation were the parties; the recognition of an *équipe nationale* made up of individual political figures now looked anomalous. Instead, the concept of an *équipe* was applied to cases where there were several Christian Democratic parties in one country that had to unite as a national *équipe* to belong to the EUCD. That suited the German parties – the CDU and the Bavarian Christian Social Union (Christlich-Soziale Union, CSU) – and even more the three Dutch Christian parties – the KVP, CHU and ARP – despite their being rivals in domestic politics.

It was again the French, and only the French – the Belgians having accepted the new order in the early 1960s – who could not decide whether their party should join the EUCD. The MRP's successor party, the Centre démocrate, led by Jean Lecanuet, wanted to be an 'open' party. And in fact, prominent figures from the Liberal camp (though progressively fewer of them) had joined. This fact excluded the Centre démocrate from becoming part of the Christian Democratic family of parties. Until 1976, the French *équipe* was led by Senate President Alain Poher; in the same year, the CDS was founded and immediately joined the EUCD.

Over the first decade of the EUCD's existence, the following parties were members: the German CDU and the Bavarian CSU, the Italian Christian Democracy (Democrazia Cristiana, DC), the Austrian People's Party (Österreichische Volkspartei, ÖVP), the Swiss Christian Democratic People's Party (Christlich-Demokratische Volkspartei der Schweiz/Parti démocrate-chrétien suisse, CVP/PDC), the CVP and the PSC from Belgium, the Luxembourg Christian-Social People's Party (Chrëschtlech-Sozial Volekspartei, CSV), the Christian Democratic Party from San Marino (Partito Democratico Cristiano Sammarinese, PDCS), the Maltese National Party (Partit Nazzjonalista, PN), and the Dutch KVP, CHU and ARP.[46] In 1972 a Spanish

---

[45]   On the development of the EUCD, see Pierre Letamendia, 'L'Union européenne démocrate chrétienne', in Portelli and Jansen, *La Démocratie chrétienne*, 55–63. See also Papini, *L'Internationale*, 94ff.

[46]   In 1957 the Swiss Catholic Conservative Party (Katholisch-Konservative Partei der Schweiz) changed its name to the Conservative-Christian-Social People's Party (Konservativ-Christlichsoziale Volkspartei) and in 1970 into

équipe consisting of Castilian Christian Democratic groups and parties was accepted into the EUCD, along with the Democratic Union from Valencia (Unión Democrática del País Valenciano, UDPV), the Catalan Democratic Union (Unión Democrática de Catalunya, UDC) and the Basque National Party (Partido Nacionalista Vasco, PNV). In 1974 the Portuguese Democratic and Social Centre (Centro Democrático e Social, CDSp) joined too.

According to Roberto Papini, a leading contributor over many years to the work of international Christian Democracy:

> There were no permanent alliances between the parties, rather natural ties, as were found in the NEI period between German-speaking parties (CDU, ÖVP, Swiss CVP/PDC), between Benelux countries, and between Italian, French, and Spanish parties. By contrast, political alliances evolve as a result of concrete problems. Beyond that, it is also worth remarking that medium-sized parties – as with medium-sized states – often evince more enthusiasm for international cooperation than many large parties, which are in a stronger position to pursue an independent foreign policy.[47]

Structure and Working Methods

The EUCD used a simple procedure for electing new members: the Political Bureau nominated from its ranks a small committee to look into applicants' principles, programmes, politics, meaning and political practice. The Political Bureau made its decision to accept new members on the basis of the report's findings and the committee's recommendations.

The constitution laid down the EUCD's tasks: deepening the Christian Democrats' guiding principles and encouraging political studies; nurturing the tradition of Christian humanism, democracy and social justice; spreading ideas and information about the achievements of Christian Democracy. It should be the aim of the EUCD 'to develop a permanent, close cooperation between Christian Democratic parties in Europe, leading to a common policy of creating a federated Europe.'[48]

Christian Democratic People's Party (Christlich-Demokratische Volkspartei der Schweiz/Parti démocrate-chrétien suisse, CVP/PDC).

[47]   Papini, *L'Internationale,* 98.

[48]   Clause in the constitution of 18 July 1971, reprinted in *Zusammenarbeit der Parteien in Westeuropa: Auf dem Wege zu einer neuen politischen Infrastruktur?* Institute for European Politics (Bonn, 1976), 332–37; this statute remained in effect until 1992, apart from a few small changes that did not alter the organisational structure.

A congress to establish political guidelines was to be convened every three years. Four EUCD Congresses were held between 1965 and 1978: in Taormina (1965), Venice (1968), Bonn (1973) and Berlin (1978); that is, twice in Italy and twice in Germany. This indicates the growing influence of DC and of the two German Union parties, especially the CDU, in international and European cooperation. The choice of personnel also reflected the dominant role of the Italians and Germans: Rumor's successor as President in 1973 was Kai-Uwe von Hassel, while Tindemans was succeeded as Secretary General by the Italian Arnaldo Forlani (1974–78).

Within the political guidelines laid down by the Congress, the Political Bureau decided on all practical and political questions, meeting at least twice a year but in practice usually three or four times annually. The Bureau also elected the president, vice-presidents and the secretary general. The Political Bureau consisted of the president, four vice-presidents, the secretary general, the president of the European Union of Young Christian Democrats (EUYCD), the chairs and two further representatives of the Christian Democratic groups in the EP and the Council of Europe (CoE), former EUCD presidents, presidents of European bodies who belonged to member parties and five delegates (originally the chair, secretary general, international secretary and two representatives of national parliamentary groups) per member party or *équipe*, independent of the strength or political importance of the group in question.

This last condition led, for instance, to the San Marino party alone having as many votes as the two German Union parties together, since they formed a single national *équipe*. It underlines the fundamentally accepted, but in practice much-criticised, confederal character of the EUCD's structure. But as circumstances evolved, the rules on representation eventually became irrelevant. The stronger, politically more important parties – whose larger financial contributions kept the organisation going – made their weight felt. Finally, cooperation in the EUCD framework did not in practice mean formulating, and certainly not implementing, practical policies with substantive content. Rather, the organisation served as a forum to exchange thoughts and information, to formulate ideas and strategies, and to forge a certain consciousness and consensus.

One substantial innovation was that the EUCD, unlike the NEI, specifically took into account the possibilities offered by the now-functioning European institutions. It was not just the representatives of national parties or national parliamentary groups who cooperated in EUCD bodies. Those who held office at the European level, especially the leading figures in the two Christian Democratic parliamentary Groups, the Assembly of the CoE and the EP, also participated. This markedly improved internal communication and the everyday vitality of the organisation. In particular, the Christian Democratic Group in the EP either took over more and more tasks from the EUCD, or made its material resources available to the EUCD.

The EUCD's resources came from member party contributions calculated according to rules laid down by the Political Bureau. In practice, however, the organisation's work and many of its activities were made possible only because of considerable extra financial support and services provided by the Italian DC, whose offices housed the EUCD Secretariat in Rome.

A Christian Democratic Doctrine

A centrepiece of EUCD activity was, from the beginning, the attempt to create a consensus about Christian Democratic doctrine; this was seen as a precondition for any joint effort to advance the cause of European union. The traditions and ideas of the member parties differed considerably, as did the spiritual and cultural climate in each of the countries in which they operated.

The Rome-based Christian Democratic International Centre for Information and Documentation (CDICID) played an important role in this connection. It had been established as early as 1960 in conjunction with the formation of the Christian Democratic World Union (CDWU). To secure financing, the Centre had been europeanised, so to speak, in 1968, and incorporated into the EUCD. It remained under the control of Karl Josef Hahn, a Dutchman, who was to do the job from the institute's inception until it folded in 1978.[49] The institute took on the task of organising ideological, philosophical and theoretical reflections and discussions about a Christian Democratic programme in tune with the times. To this end, meetings were organised of officials in charge of publications and education for the various member parties. In the course of the 1970s, the institute published *Christian Democratic Panorama*, a cultural magazine that appeared in several languages. *Cahiers d'études*, which appeared at the same time, contained numerous studies and reports. It focused on basic social

---

[49] From 1971 to 1982 Hahn was also Deputy Secretary General of the EUCD. On the CDICID, see Papini, *L'Internationale*, 100ff.

and political issues, and Christian Democratic parties' different political programmes and experiences. All these efforts were also put to the service of advancing dialogue with the Latin Americans in the framework of the CDWU. Later, Papini declared: 'The overriding importance of the Centre is that it made possible the deepening of Christian Democratic culture at an international level, attempting to reconcile European parties' positions with those from Latin America.'[50]

One important achievement in this area took place on 12 February 1976, when the EUCD Political Bureau, meeting in Paris, reached agreement on a 'Manifesto of Christian Democrats in Europe'.[51] A little later, on 16 July 1976, the CDWU's Political Bureau followed suit by passing its Political Manifesto.[52] Much of that text was drawn up by an enthusiastic CDICID working group, named Idea and Action. The Political Manifesto involved a compromise, in that it included the various elements that the parties insisted were integral to their own specific traditions: personalist-communal and Christian-social ideas, along with concepts that were Christian-conservative and pragmatic-centralist.

Between a European Community Orientation and the 'Conservative Question'

From the early 1970s onwards the EUCD was increasingly compelled to address the question of whether, and how, the Conservatives should be involved in Christian Democratic joint activities. The issue moved to centre stage after the United Kingdom joined the EC in 1973. In the 1960s the British Conservatives, no doubt anxious to avoid isolation, as well as to secure some kind of leading role, had adopted the practice of holding 'inter-party conferences'. These took place on the fringes of the party conference proper and developed into

a forum for informal debate. Usually attended by Conservative representatives from the Scandinavian countries, there were also delegates from German, Austrian and Swiss Christian Democratic parties. As the EC took shape and its membership grew, British and Scandinavian Conservatives and Christian Democrats from the neutral countries felt a greater need for a more intensive flow of information and

---

[50]   Ibid. 101, n. 169.
[51]   Text in *Programm und Statut: Dokumentation 1*, European People's Party, 2nd ed. (Brussels, 1984), 23–31.
[52]   Ibid. 33–41.

cooperation. The idea of establishing a grouping of parties making up the broad European centre had been mooted for quite some time.[53]

The EUCD accordingly began to debate whether a 'democratic centre' should or could be created.

As all these developments were put into progress, it became apparent that the European party system would have to undergo radical changes. Spurred on by the continuing integration process, parties in EC countries were paying more attention to EC-related policies and institutional and procedural matters. Proceedings in EUCD bodies were increasingly overshadowed by the EP's Christian Democratic Group, or by the parties represented in it.

EUCD member parties from non-Community countries, and in particular the neutral Austrians and Swiss, increasingly felt their interests were being neglected and their views relegated to the sidelines. The Standing Conference of the Six, set up within the EUCD in 1970, failed to allay such anxieties, as did the Political Committee of Christian Democratic Parties from the EC Countries, which was established two years later. However, both made for clarity and a rational division of tasks. The precarious situation brought about by these developments was further exacerbated when the EPP was founded in 1976 and a crisis within the EUCD followed. It was a crisis destined to continue until the watershed year of 1989, when the collapse of Communism opened up a new field of activity in Central Europe and the EUCD had a new *raison d'être*.

These two problems – the turbulence produced by rapid progress towards EC integration, and relations with the Conservatives – sorely tested the EUCD's consensus. There was an incipient awareness of the need to form a 'European party'; it grew more and more evident as the Community's distinctive political system gradually emerged.

---

[53]   Heinrich Böx, 'Demokratie im christlichen Europa', in Philipp Jenninger et al. (eds.), *Unverdrossen für Europa: Festschrift für Kai-Uwe von Hassel* (Baden-Baden, 1988), 144.

Chapter Two

The Founding of the EPP

The first steps towards creating the European People's Party (EPP) were taken by the Christian Democratic Group in the European Parliament (EP). In order to meet the challenge of establishing a European party organisation in the run-up to the first direct elections of the EP, a Political Committee was established within the European Union of Christian Democrats (EUCD). It functioned as a bridge between the parliamentary groups and the parties of the Christian Democratic family. It was within that committee that decisions were taken about the programme, the internal organisation and, not without dispute, the name 'European People's Party'. The ambition and first realisations of this new party federation went beyond similar initiatives of other European parties, but were overshadowed by the founding of the European Democratic Union (EDU).[54]

Building Parliamentary Groups in the European Assemblies

The first European assembly created after the Second World War was the Council of Europe (CoE). Its Consultative Assembly was founded in 1949, only four years after the end of the war. However, it was not until much later, in the early 1960s, that like-minded parliamentarians from the different member countries united in groups. One reason was that the Con-

---

[54] On the founding of the EPP, see also Wilfried Martens, *Europe: I Struggle, I Overcome* (Brussels, 2009), 35–42 and Steven Van Hecke, 'On the Road towards Transnational Parties in Europe: Why and How the European People's Party Was Founded', *European View* 3 (Spring 2006), 153–60.

sultative Assembly met infrequently, and cooperation imposed few obligations on anyone. Besides, the CoE's role in matters such as democratic development, the rule of law and human rights was not something about which traditional political families held opposing views, and neither did the founding Member States and their representatives. The incentives for close cooperation along party lines were insufficient or simply nonexistent.

Initially, that kind of close cooperation was not expected in the General Assembly of the European Coal and Steel Community (ECSC), either. The basic treaties of the European Communities distributed seats to parliamentarians exclusively by Member State. The mandate of those deputies, who were nominated by their national parliaments, came from voters in Member States. Accordingly, it was assumed that the General Assembly would in practice divide along national lines. The grouping together of parliamentarians by political groups in the General Assembly in 1952 was therefore rather revolutionary. Given the ECSC's specific competences, as well as the debate about the European Defence Community (EDC), opinions within the different Member States differed considerably. Much more affinity and understanding was found within the political families.

The establishment of parliamentary groups in the ECSC General Assembly was formally recognised in a resolution of 16 June 1953.[55] The first official meeting of the Christian Democratic Group took place on that day. The declaration that created it was lodged on 23 June. Christian Democrats had gathered several times since their first meeting on 11 September 1952, one day after the Assembly's inaugural sitting.[56] The Group was composed of 38 members out of a total of 78 Assembly deputies. When in 1958 the General Assembly also became the parliamentary organ of the European

---

[55] Published in the *Official Journal of the European Coal and Steel Community*, 21 July 1953. An additional clause in the rules of procedure allowed for the legal possibility of establishing parliamentary groups and made this formally dependent on a declaration that such a group was being founded, which in turn required a quorum of nine members (about 12%).

[56] Pascal Fontaine, *Voyage to the Heart of Europe (1953–2009): A History of the Christian-Democratic Group and the Group of the European People's Party in the European Parliament* (Brussels, 2009), 43; Hans M. ten Napel, 'Van het continentale naar het angelsaksiche model van christen-democratie? Over de problematische europeanisering van de christen-democratische politiek', in *Jaarboek Documentatiecentrum Nederlandse Politieke Partijen* (Groningen, 1997), 233.

Economic Community (EEC) and the European Atomic Energy Community (EAEC), it changed its name into 'European Parliament' (EP). Article 36 of the rules of procedure explicitly included the right to form parliamentary groups. The Christian Democratic Group, which then consisted of 66 deputies, stated in its own rules of procedure that membership was open to those belonging to a Christian Democratic party (according to the group's definition of 'Christian Democratic' and the party's values).

In turn, the EP fostered an increasing desire to improve inter-party cooperation. The national parties that had come together at the European level did not automatically pursue the same policies, even if they agreed ideologically. The constant effort to find common positions led to the discovery of common fundamentals and a recognition of their importance to common action. It also led to better information, increased understanding and finally to an appreciation of how different the parties were from each other. Joint debates with opponents or competitors encouraged the feeling of affinity. For the first time, the experience of belonging to a family of parties was not only the privilege of the party leadership but could also be shared by parliamentarians.

At the same time, European parliamentarians were exposed to the danger of remaining isolated in Strasbourg and Luxembourg. Their dual mandate, which meant they had to be present at both national and European levels, placed a huge burden on everyone, in intellectual, practical and physical terms. When something important was going on in a national parliament, the member might often be absent from the EP. For the most part, national problems overshadowed European issues because they were more urgent and because they directly impinged on a deputy's position in his or her constituency, party or parliamentary group. Colleagues in national parliamentary groups or parties were rarely in a position to deal with the complexity or range of European problems. Dialogue between European parliamentarians and their national parties and groups was the exception, not the rule.

The Christian Democratic structures were not yet adapted to these new developments and challenges. The Nouvelles équipes internationales (NEI), founded at a time when there was not yet talk of a European assembly, lacked a connection with the Christian Democratic Group in the EP. What made things worse was that inside the NEI, there was much less cooperation between the parties from the mid-1950s onwards. Therefore, the Christian Democratic Group found itself in a very unsatisfactory and ineffective position. It had no organic connection with an appropriate party organisation to support and encourage it. Its members and leadership had to operate without any solid or clear organisational framework from the party family.

The creation of the EUCD in the mid-1960s completely changed that situation. The EUCD made increasing efforts to support the group in the EP in facing up to the new tasks of the European Community (EC) as it developed and grew in importance.[57] Close contact with the EUCD was maintained through Alain Poher, President of the EP from 1966 to 1969, and Hans August Lücker, Chair of the Christian Democratic Group from 1969 to 1975. At the same time, EUCD bodies were being properly established by its President, Mariano Rumor. Cooperation between the parliamentary group and the EUCD was eventually given institutional form with a new statute on 18 July 1971.

Linking the Groups with the Parties: The Political Committee

The constant improvement of integration within the EC and the simultaneous and growing need to reach agreements between Member States meant there had to be closer cooperation between their representatives. The next development, in April 1970, was an informal conference of leading figures from EUCD member parties from EC countries. In April 1972, a special body was set up in the context of the EUCD: the Political Committee of Christian Democratic Parties from Member States of the European Communities. Its goal was to establish 'a permanent link between the parties and parliamentary groups at national and European levels, as well as a basic political consensus about the deepening and further development of European integration'.[58] According to Arnaldo Ferragni, the Secretary

---

[57] See Karl Josef Hahn and Friedrich Fugmann, 'Die Europäische Christlich-Demokratische Union zwischen europäischem Anspruch und nationalen Realitäten', in Karl Josef Hahn (ed.), *Zusammenarbeit der Parteien in Westeuropa: Auf dem Weg zu einer neuen politischen Infrastruktur?* Institute for European Politics (Bonn, 1976), 304–31.

[58] Article 1 of the by-laws of the Political Committee of Christian Democratic Parties from the Member States of the European Communities, 7 April 1972, reproduced in *Zusammenarbeit der Parteien*, 338ff.

General of the Christian Democratic Group, the Political Committee's task was to develop 'an organisational structure . . . appropriate to use as a "structure d'accueil" for a possible future European Christian Democratic party'.[59]

The presidents of the EUCD and the Christian Democratic Group co-chaired the executive of the Political Committee. This consisted of leading figures from parties in the EC and members of the Group's executive in the EP. Soon there was more and more contact and cooperation, which in turn encouraged unified political action. Joint conferences of the Christian Democratic Group and representatives of national Christian Democratic parliamentary groups were organised. Working groups were set up involving a mixture of representatives of the Christian Democratic Group, EUCD bodies and/or member parties. Together they elaborated political programmes, forged a consensus in every area relevant to European policy and clarified the main points of Christian Democratic ideology.

The leader of the Christian Democratic Group in the EP was ex officio on the EUCD executive committee as well. This offered the permanent possibility of consulting the EUCD leadership and of maintaining contact with chairs of national parties and parliamentary groups, and with Christian Democratic heads of government and ministers. Lücker was adept at exploiting all these possibilities to achieve the breakthrough formation of a 'European party'. This had been his goal, shared by a number of others.[60] On his re-election as chair of the Christian Democratic Group in 1974, Lücker assured his audience 'that in the course of his new mandate he would continue to devote himself wholeheartedly to the interests of the parliamentary group and the European Union of Christian Democrats so that a European Christian Democratic Party could be founded, a party with a single political programme'.[61]

---

[59] Letter from Ferragni, Secretary General of the Christian Democratic Group, to Karl Josef Hahn, Deputy Secretary General of the EUCD, 4 January 1972, IX-004-081.

[60] See articles by Egon Klepsch (pp. 31–33), Ferragni (pp. 84–88) and Josef Müller (pp. 111–13) in Fraktion der Europäischen Volkspartei, *Liber Amicorum: Erinnerungen an Hans August Lücker zum 70. Geburtstag* (Bonn, 1985).

[61] Minutes of the meeting of 12 June 1974 in Strasbourg, ACDP IX-001-008/1.

The organic cooperation between the Christian Democratic Group and the EUCD was deepened and intensified during the presidency of Kai-Uwe von Hassel, who in 1973 succeeded Rumor as leader of the EUCD. However, the EUCD was still a relatively loose association of parties, many of them from non-EC countries. The founding of the Political Committee had already underlined the fact that the EUCD was not an adequate organisation as far as the promoters of a true European federation were concerned.[62] Members of the EP were particularly dissatisfied. They sorely missed an active party organisation at the Community level.

Direct elections to the EP were being discussed, and the EC's political system was taking shape. The heads of state and government, at their meeting in Paris in 1972, had given notice that they intended to 'transform the totality of relations between Member States . . . into a European Union before the end of this century'. The Community admitted three new members on 1 January 1973: the United Kingdom, Ireland and Denmark. European Political Cooperation (EPC) was put in place, and discussion began on establishing an economic and currency union. A new EC financial statute was to be agreed upon, and in this connection, too, it was argued that there should be 'no federal treasury without federation'; in other words, any reforms should give Community institutions a federal character.

The Prospect of Direct Elections and European Parties

As early as 1970, during a study conference of European deputies from the Dutch Christian Democratic *équipe*, Tjerk Westerterp had recommended the founding of a European Christian Democratic party as both possible and necessary.[63] He explained that European parties were needed for much

---

[62] Walter Hallstein, former President of the EEC Commission (1958–68), published a book about his experiences and the political philosophy underlying his work in the Community, under the significant title *Der Unvollendete Bundesstaat* [The Unfinished Federal State] (Düsseldorf, 1969). It was translated into English by Charles Roetter and published as *Europe in the Making* (London, 1972).

[63] The Dutch parties represented in the Christian Democratic Group – the Catholic People's Party (Katholieke Volkspartij, KVP), the Christian Historical Union (Christelijk-Historische Unie, CHU) and the Anti-Revolutionary Party (Anti-Revolutionaire Partij, ARP) –, who were competitors at home and who generally did not follow the same political line, worked together more and more after

the same reason as national parties, to enable citizens to participate, to formulate alternatives, to choose candidates and to serve as a channel for new ideas: 'All pre-conditions for the foundation of European parties will have been met at the latest by the time of the general election of the European Parliament in all Member States.'[64]

In the Dutch parliament a few days before the 1970 study conference, Westerterp had proposed the prompt introduction of direct elections for national deputies to the EP, as provided for in the Community treaties. The matter was becoming ever more urgent, both for European and for democratic reasons, he said. Soon afterwards, a parallel initiative for direct election of German European deputies was proposed by several Bundestag parliamentarians from the German Christian Democrats (Christlich Demokratische Union/Christlich-Soziale Union, CDU/CSU). In several other countries, there were similar attempts to ensure that direct elections to the EP should follow national electoral laws until such time as European regulations had been agreed upon. There was evidently another motive for all these proposals: to put pressure on governments to get moving on the issue of direct elections and democratisation.[65]

On 1 January 1973, Lücker, as Chair of the Christian Democratic Group, sent EUCD President Rumor a note concerning 'the political activity of the CD [Christian Democratic] Political Committee of the European Community in connection with the decisions of the October 1972 Paris Summit'. In this note, he demanded

an intensification of cooperation in the Political Committee as well as in the Christian Democratic Group's communication with groups in the national parliaments . . . But this alone will not be enough. Such work must be pursued more systematically, make a far more effective impact on the activities of national parties and parliamentary groups, and eventually be carried into popular political consciousness. This is impossible without appropriate personnel and financial re-

---

1967, and as a result of their common membership in the EUCD established a 'contact committee' on 17 June 1972. A strategy paper was presented to this committee calling for 'the formation of a single party'. This project was realised in 1977 with the founding of the Christian Democratic Appeal (Christen Democratisch Appèl, CDA). See ACDP IX-004-100/8.

[64] The text was published in various languages by the Christian Democratic Group of the European Parliament in *CD-Europa* (published by the Christian Democratic Group in the EP, Luxembourg) on 16 July 1970. See ACDP IX-001-055/2.

[65] See Karl-Heinz Nassmacher, *Demokratisierung der Europäischen Gemeinschaften* (Bonn, 1972).

sources. It is true that the existing organisational structures can be used, but not that they take on the whole of this new task, for which the organisation was not intended, and for which it is not equipped. [A political programme has to be elaborated] whose declarations on all the important areas in political life can be jointly presented by parties across the Community as their joint election manifesto for the direct elections to the European Parliament. [To get through the work needed to prepare for the European elections] the Political Committee must engage its own secretary-general, based in Brussels. The rule that the parliamentary group's secretary-general should double as secretary-general of the Political Committee is also less and less sustainable: increasing demands are being made of the parliamentary group. [The Christian Democratic Group also] intends to create a political institute in the form of a foundation, which might be given the honorary designation of, for instance, the Robert Schuman Foundation.[66]

The idea of giving the Political Committee its own secretary general in order to take a further step towards independence and the creation of a European party was not put into practice. The move was vetoed by the parliamentary group's Secretary General, Alfredo De Poi, who saw a threat to his own political function and a risk of damaging coordination between the two bodies. All in all, Lücker's proposals proved premature. But they did show the way things were intended to develop, even if the realisation of those plans was to be delayed for a while.

It was indeed 'already in the logic of things that Christian Democrats were thinking of building a European Christian Democratic party – it emerged from all the forms and processes of their European cooperation . . . and the external impulse for this' would be 'the direct elections to the European Parliament'.[67] From the start, however, there was not enough agreement about how to proceed with constructing this European party, and particularly about its political and ideological profile. The European Union of Young Christian Democrats (EUYCD), for instance, took the revolutionary position that national Christian Democratic parties should be replaced by 'one large Christian Democratic party'. This emerged from a declaration made at their meeting in Malta on 15 May 1972, in which they proposed a 'party which will give popular expression to political participation in the framework of free, communal institutions; it should be marked by a progressive spirit and find coherence in the anti-fascist tradition. This

[66] ACDP IX-004-082 [translated from the 1998 version of the book].
[67] Hahn and Fugman, 'Die Europäische Christlich-Demokratische Union', 329.

party will solve current problems which national political forces cannot deal with because they are too small.'[68]

The dominant doctrine, by contrast, was expressed in a contribution by Robert Houben, head of the Belgian member party's research institute, the Centre for Economic, Political and Social Studies (Centrum voor Economische, Politieke en Sociale Studies, CEPESS):

> Both political integration of the kind we are striving for and the fact that the European Parliament exists pose the question of forming parties at European level ... At the same time, to be realistic, we must assume that a party at European level will only take shape gradually, and in fact will start from what exists; just as European union can probably only come about through the development of the Common Market to an Economic and Currency Union, and then on to political integration, which will have to take into account a transition period and a great deal of patience ... Building a party at European level and the further development of European integration have to develop in parallel and at the same pace.[69]

The final incentive to realise this project came from the EC's heads of state and government, who decided in December 1974 to fix the date of the first direct elections to the EP for 1978.[70] The Political Committee of Christian Democratic Parties from European Community Member States began talking about the practical preparations for these elections as early as spring 1975. Various ideas were discussed. In Bonn, for instance, acting on behalf of EUCD President von Hassel, Heinrich Böx, head of the CDU's office for foreign relations, headed up a small study group.[71] With the founding of a European party in its sights, the group discussed ques-

---

68 'Dossier sur la formation d'un parti européen', Doc 16/10.9.1975, ACDP IX-004-096 [translated from the 1998 version of the book].

69 Robert Houben, 'La formation d'un parti européen', ACDP IX-004-096 [translated from the 1998 version of the book].

70 As there was a lot of delay in the preparation for the first direct elections, the European Council finally decided at its meeting in Copenhagen (7–8 April 1978) to fix the dates at 7–10 June 1979.

71 The group consisted of Meinhard Ade (CDU planning staff), Hans Herbert Holzamer (CDU department of foreign relations), Thomas Jansen (personal representative of the then CDU party and CDU/CSU parliamentary Group Chair Rainer Barzel MP) and Hubertus Dessloch and N. Obermeier (both CSU members and senior officials in the Bavarian Land Representation). See Eva-Rose Karnofski, *Parteienbünde vor der Europa-Wahl 1979: Integration durch gemeinsame Wahlaussagen?* Institute for European Politics, vol. 59 (Bonn, 1982), 191ff.; and Thomas Jansen, in Fraktion der Europäischen Volkspartei, *Liber Amicorum*, 115.

tions of strategy, a political programme and an organisational structure, and prepared documents for the German delegation in the Political Committee. This group also proposed the name European People's Party.

Establishing a European Party

In September 1975, an ad hoc European party working group was set up and chaired jointly by Wilfried Martens, President of Belgium's Flemish Christian People's Party (Christelijke Volkspartij, CVP), and Lücker, Chair of the Christian Democratic Group in the EP.[72] Members were the Deputy Secretary Generals of the EUCD, Karl Josef Hahn (Netherlands) and Böx (Germany); the Secretary General of the Christian Democratic Group, De Poi (Italy); and the Executive Secretary of the EUCD, Josef Müller (Germany). The working group was given the task of 'elaborating the basic documents for founding a Christian Democratic party with reference to direct elections to the European Parliament', in other words, a constitution and a programme. The working group met several times between November 1975 and January 1976, and in Paris, on 20 February 1976, was able to present the Political Committee with a draft constitution.[73] Having checked the draft, which was rapidly accepted, the Political Committee meeting in Brussels on 29 April 1976 decided to found the European People's Party: Federation of Christian Democratic Parties of the European Community, and adopted the constitution.[74]

The formal founding of the EPP took place in the context of a meeting of the EUCD Political Bureau in Luxembourg on 8 July 1976, in fact, the first and constitutive meeting of the Political Bureau of the EPP.[75] Representatives of the following parties took part: the Flemish CVP and the French-speaking Christian Social Party (Parti social chrétien, PSC) from Belgium; the CDU and CSU from Germany; the Democratic and Social Centre (Centre des démocrates sociaux, CDS) from France; the Family of the Irish (Fine Gael, FG) from Ireland; Christian Democracy (Democrazia

---

[72] Decision of the Political Committee meeting, Luxembourg, 26 September 1975.

[73] ACDP, KAS IX-004-096.

[74] Published in *Programm und Statut*, EPP Dokumentation Series, vol. 1, 2nd ed. (Brussels, 1984), 45ff.

[75] Documented in *CD–Europa* 2 (1978).

Cristiana, DC) from Italy; the Christian-Social People's Party (Chrëscht-lech-Sozial Vollekspartei, CSV) from Luxembourg; and the Catholic People's Party (Katholieke Volkspartij, KVP), the Anti-Revolutionary Party (Anti-Revolutionaire Partij, ARP) and the Christian Historical Union (Christelijk-Historische Unie, CHU) from the Netherlands.

Leo Tindemans, Prime Minister of Belgium and former Secretary General of the EUCD, was unanimously elected President. In accordance with the rules of procedure, the EUCD President, von Hassel, and the Chair of the Christian Democratic Group in the EP, Alfred Bertrand, became ex officio Vice-Presidents of the EPP. Further Vice-Presidents were elected: Norbert Schmelzer (Netherlands), André Colin (France) and Dario Antoniozzi (Italy). Former Bundestag and EP member Müller, who had been the EUCD's Executive Secretary in Brussels since 1973 when the EUCD General Secretariat was in Rome, was given the task of building up the EPP General Secretariat. He would, for the time being, run things in tandem with the Group's Secretary General, Giampaolo Bettamio.

The process of founding the party also involved agreement on the political programme at the EPP's first Congress in Brussels on 6–7 March 1978. The proposal elaborated by rapporteurs Martens and Lücker was accepted without amendment: it had already been agreed on by the EPP's Political Bureau, which had discussed it several times.[76] On the margins of the Congress, at the initiative of the President, the EPP's Political Bureau also elected a Secretary General, Jean Seitlinger. A member of the French National Assembly, in the early 1960s he had served as the last secretary general of the NEI. Bertrand had meanwhile been succeeded as Group Chair by Egon Klepsch and was elected Treasurer, completing the leadership. One week later, on 14 March, the Christian Democrats in the EP added to its name Group of the European People's Party. The structure of the EPP was now complete.

---

[76] Published in *Programm und Statut,* 6ff. See also the following two publications, which provide further details about the development of the programme, with comments and explanations: Karnofski, *Parteienbünde,* 387ff.; Martin Bangemann, Roland Bieber, Egon Klepsch and Horst Seefeld, *Programme für Europa: Die Programme der europäischen Parteibünde zur Europa-Wahl 1979* (Bonn, 1978), 181ff.

Controversy over the Name

Agreement had quickly been reached about the structure of the new party and how it would be built up. From the outset, there was consensus about the goal – the creation of a political party. But naming the organisation was more difficult. The draft constitution put before the Political Committee in February 1976 had not resolved the issue. The name had not yet been discussed at this point, and it remained contentious, as it would define the party's ideology, alignment and political profile. A name signals not only the message to be conveyed; it also gives an idea of who is delivering the message and whom the message is for.

The row was set off by the strategic question of whether or not the British and Danish Conservative parties, whose deputies in the EP had formed their own group, should be invited to join the future European People's Party.[77] Those in favour of the invitation argued for a name that avoided the description 'Christian Democratic', which they felt was too narrow and exclusive. Those against the invitation – in other words, those who wanted to give the new party an unambiguously Christian Democratic character – argued that the character should be reflected in the name. In that way, there would be no possibility of confusion.

Above all, the Germans favoured a 'strategy of openness'. Their arguments were evidently based on their own experience and tradition. Both the CDU and CSU owed their own strength, relative to affiliated parties elsewhere in Europe, to being able to integrate Conservative (and Liberal) forces. Against them, the Italian DC, along with the Dutch and Belgian parties, insisted on emphasising a Christian Democratic identity. They saw this as important to the party's *raison d'être* and political direction, and essential if their action in helping to found a European party were to be successful in their own countries. The party's identity, its political profile and therefore its whole coherence and effectiveness would, it was argued, be hopelessly compromised if the Conservatives were allowed to join at this crucial early stage. The other founding parties from Luxembourg, Ireland and France agreed.[78]

---

[77] In 1973 a Conservative Group was established. In 1979 it was renamed into European Democratic Group (EDG). It ceased to exist in 1992, following the entry of British and Danish Conservatives in the EPP Group. See chapter three of the present volume.

[78] The case against the pro-Conservative position is argued by P. H. Kooijmans in 'La Démocratie chrétienne en Europe' (translated from a contribution to 'Anti-

To reach a compromise with the Germans, it was agreed that soundings would be taken among other like-minded parties in order to establish a common platform, the 'Democratic Centre committed to European unity and social progress'. This was not meant to be a party:

What we had in mind was, rather, an association like a political club, to provide the framework for meetings and discussions between the leading political figures in member parties and parliamentary groups. The final impulse for these ideas was the political alliance of the Socialists with the French Communist Party, as well as the development of some EC Communist parties into 'Euro-Communism'. These were plainly efforts to achieve a joint political structure involving both Socialist and Social Democratic parties.[79]

Under no circumstances was such an initiative to be pursued at the cost of the effectiveness of the European party of Christian Democrats. In September 1975, following a proposal by Lücker, a working group with the name 'Democratic Centre' was set up alongside the European Party working group. The efforts of the new working group, however, were fruitless.

Ultimately, there was a clear majority against bringing in the Conservatives, but as far as the name was concerned, a compromise was found. The idea of a 'People's Party' signalled both the openness the Germans wanted and a connection to the Christian Democratic tradition that was important to the others. Numerous past and present Christian-inspired parties and parties with a Christian Democratic or Christian Social orientation had similar names. For example (to mention only parties in the then EC Member States), there was the Italian People's Party (Partito Populare Italiano, PPI); and in France, the Popular Democratic Party (Parti démocrate populaire, Pdp) and the Republican People's Movement (Mouvement républicain populaire, MRP); the Belgian CVP; the Dutch KPV and the CSV in Luxembourg. To safeguard the new party's identity and guarantee it, the official name European People's Party was followed by a descriptive subtitle: Federation of Christian Democratic Parties of the European Community. This subtitle was, however, rarely used and consequently soon forgotten.[80]

---

revolutionaire Staatskunde', Den Haag, March 1973), in ACDP IX-004-096.

[79] Hans August Lücker and Karl Josef Hahn, *Christliche Demokraten bauen Europa* (Bonn, 1987), 130.

[80] Note that at the time the NEI was founded, the same idea – avoiding the term 'Christian Democracy' – was very strong and supported by the French (with reference to the French Christian Democratic parties that had never had 'Christian Democracy' in their names). See chapter one of the present volume and

A Federative and Open Party Organisation

Apart from its character of compromise, the name expressed a bold challenge and an honest judgement by the party's founders. It was bold because they were aiming high by calling the EPP a party and setting a goal whose realisation seemed to many, at the time, a chimera. It was honest in that the subtitle expressed what it was really about, namely an association of parties. There was a dynamic tension between that forward-looking claim to being a European party 'in the making' and the realistic insight that it could for the moment be no more than a European association of national parties. And it was out of that fruitful tension that, over the coming years, the EPP was to develop.

As a European party, the EPP could not – and did not want to – copy any national model. Unlike national parties, no single model informs every part of its organisation. The EPP respects the existing, established and valued structures of its member parties. It builds on them and depends on them. In other words, the EPP was founded as a federative party which organises common action by its members at European level and seeks to ensure they are politically supported. This orientation is also expressed in the party's constitution.

Two categories of membership were foreseen for the various party bodies, such as the Executive Committee, the Political Bureau and the Congress. One sort of member would be the representatives of member parties or, where there was more than one member party in a country, the national *équipe*. Their number would be determined by the proportion of national deputies in the EPP Group. The other category would consist of elected deputies or office holders belonging to member parties in the EP, the European Commission, the EPP Group and associations recognised by the EPP, as well as, obviously, the president of the EPP itself, its vice-presidents, treasurer and secretary general. Every decision would require an absolute majority of members present, a rule which reflected the party's supranational and democratic character.

The Political Programme

The political programme decided by the EPP's first Congress in 1978 evinced considerable consensus on almost all European political issues,

---

Van Hecke, 'On the Road towards Transnational Parties in Europe', 153.

many of them with profound socio-political implications. The programme expressed a joint intention to develop and complete integration in the context of the EC, leading to a political union equipped with federal and democratic institutions.

The rapporteurs, Martens and Lücker, had been able to base their draft on productive preparatory work and agreed, documented common positions. The years prior to the founding of the EPP had been fruitful for the Christian Democrats in terms of developing political programmes. Nearly all European parties had discussed both basic and action programmes, and had either decided on them or were in the process of doing so. The reasons can be traced, above all, to the general change in the socio-political climate resulting from the revolt of 1968. This especially affected Christian Democratic parties, because those parties had generally been in government since the war. At the same time, new questions were being raised by developments in European politics and their cultural, economic and socio-political implications.

Beyond that, domestic circumstances, in Germany and the Netherlands, for instance, necessitated intensive – and extensive – efforts. In Germany, where the CDU and CSU had been in opposition since the autumn of 1969, it was a matter of a radical new direction. In the Netherlands, the three Christian confessional parties (ARP, CHU and KVP), looking ahead to uniting as one Christian Democratic party, had conducted an intensive debate about the inspiration and contents of their policy.[81]

All these multifaceted labours yielded the raw material for the rapporteurs and for everyone else involved in developing the policies of the EPP or participating in the debate. This raw material included specialised programmes, as well as important statements on key issues, election programmes, and – not least – work done by the Christian Democratic Group in the EP. Thanks to the work of the Konrad Adenauer Foundation, synopses and comparative analyses were available.[82] The 'Manifesto of European Christian Democrats', adopted by the EUCD Political Bureau in

---

[81] 'Plichten en perspectieven. Aanzet voor een profielschets van het CDA', Centrum voor Staatskundige Vorming (The Hague, 1975).

[82] Eggert von Petersdorff (ed.), *Synopse der Parteiprogramme der Christlich-Demokratischen Parteien Westeuropas*, Politische Akademie Eichholz (Wesseling, 1971). See also Bernhard Gebauer, *Die europäischen Parteien der Mitte: Analysen und Dokumente zur Programmatik christlich-demokratischer und konservativer Parteien Westeuropas*, Handbücher der Politischen Akademie Eichholz, vol. 6 (Bonn, 1978).

Paris on 21 February 1976, served as the basic document. The 'Political Manifesto of the World Union of Christian Democrats' was also helpful. This had been passed by the EUCD Political Bureau in Rome on 16 July 1976. Both texts, which Lücker had personally helped to draft, contained detailed expositions of basic precepts of Christian Democratic policy; they also set out the basic conceptual framework.[83]

The EPP Political Programme declarations were arranged in five chapters under the slogan 'Together for a Europe of Free People'. The chapters were entitled 'Our Guidelines for Europe', 'Europe in the World', 'European Community Policy', 'The Community's Institutional Framework' and 'Our Goal: A United Europe'.[84] Starting from a consensus on principles about the individual and society, the chapters contained an action programme that, given that the exercise had never been tried before, was already unusually concrete and detailed.[85]

Compared with equivalent texts passed at the same time by the European organisations of the Social Democratic and Liberal parties, the EPP's Political Programme was unique in its attempt to respond fully and in detail to relevant issues. The Social Democrats' 'Appeal to the Electorate', approved by the federation's congress in Brussels in January 1979, 'does not go ... beyond a brief presentation of global aims'. As for the Liberals' election programme, voted through by their federation's congress in Brussels in November 1977, 'only general theses [are] binding ... Statements offering detailed proposals and solutions to political problems are, as far as the parties are concerned, optional and subject to interpretation.'[86]

The EUCD Crisis

The German CDU and CSU delegates had failed to win adequate support from other parties for 'opening up' the EPP, though they had reached a compromise on the question of the party's name. But this was not the end of the matter. The Germans sought other ways to unite centre forces in

---

[83]   Texts published in *Programm und Statut*, 27ff. and 33ff.

[84]   See the first annex of this volume.

[85]   The story of the EPP's conception, founding and content are set out in detail in Egon Klepsch, 'Das Programm der EVP', in Bangemann et al., *Programme für Europa,* 77ff.

[86]   Karnofski, *Parteienbünde*, 245, 248.

Europe. Within both the EUCD and the EPP, they called for a dialogue to be established with Conservative parties, reactivating the 'democratic centre' idea. A working group to foster such a dialogue held two meetings, in December 1977 and April 1978, chaired by the EUCD President (and EPP Vice-President), von Hassel. A draft dialogue paper submitted by the CDU/CSU delegation stipulated that member parties had to pledge primary support to the EPP, since this was the only Europe-wide federation of parties to which they belonged. It was proposed that a working group be set up in parallel to the EPP to provide a platform for cooperation among all non-socialist, anti-collectivist, centre-right parties across the whole of Europe.[87]

Once the EPP had been founded, a degree of pressure to establish relatively formal links between Christian Democratic and Conservative forces was also exerted by EUCD parties that felt excluded from the EPP. An explicit stipulation had been made in the EPP constitution to the effect that only parties from EC Member States were eligible to join. On 28 October 1976, the EPP's Political Bureau decided that no other parties should be granted associate membership or even observer status. In this matter, too, the parties that set particular store by their Christian Democratic identity proved to be the most determined. They were against any dilution of the supranational and federalist approach that had inspired the formation of the EPP as a party designed to operate in the EC system.

The Austrian People's Party (Österreichische Volkspartei, ÖVP) and the Swiss Christian Democratic People's Party (Christlich Demokratische Volkspartei der Schweiz/Parti démocrate-chrétien suisse, CVP/PDC) responded by drafting a joint memorandum on 23 November 1976.[88] They wanted a revision of rules and decisions that they regarded as discriminatory, and they wanted a mechanism allowing EUCD parties to work with the EPP. Their bid, however, was turned down. Apart from the negative psychological effects this had, there was the practical consequence that the EUCD developed into a 'waiting room club'.[89] The result was that some of those affected – not only the Christian Democratic parties from Austria and Switzerland but also those from Portugal, Spain and Malta – felt the need to look for an alternative. They found it in the EDU, whose champions commended it as an alternative: 'The downgrading of the EUCD to a

---

[87] ACDP IX-004-095.
[88] ACDP IX-004-096.
[89] Kai-Uwe von Hassel, note to CDU Chair Helmut Kohl, n.d., ibid.

rump could be averted by the creation of the EDU, which would include Conservative and other centre parties alike. The EUCD could operate under the EDU umbrella as a grouping of Christian Democratic parties.'[90] The awkward result was that the hard insistence on a federal European ideal helped ensure that the EDU model, the very model the opponents to Conservative involvement were trying to resist, had a real chance of success.

The Founding of the EDU

The EDU project of developing closer cooperation between Christian Democrats, Conservatives and other 'like-minded parties' was in the tradition of so-called inter-party conferences. These had convened regularly since the 1960s, informally bringing together leading personalities from interested parties. The ÖVP was particularly keen on fleshing out this idea and creating an organisation; it was to play a leading part both in founding the EDU and in developing it. The German parties were very enthusiastic too, as was the British Conservative Party.

On the basis of their own experience, the Germans were trying to 'normalise' things at the European level. No doubt they thought it utterly perverse that they were not able to work together in the same grouping as their natural partners, the British Conservatives. The Tories were a party from a Member State which was a key player in both European policy and the 'German Question'. The principal British goal was to break out of its isolation, but in the end, Britain, too, was looking for a new order, one consonant with the adversarial nature of their system of government – an order that would bring anti-socialist forces together. A resolution at the 1975 British Conservative Party Conference called for the party to cultivate European allies: 'Recognizing that the United Kingdom is now a permanent member of the European Community . . . [the party] should work more closely with our political allies in Europe with a view to forming a moderate centre-right alliance (a "European Democratic Party") which

---

[90]   Christoph Brüse, official in the CDU foreign relations department, report on the meeting of EPP/EUCD bodies in Brussels, 20–23 October 1976.

could effectively counter Socialist coalition tactics in the European Parliament, and take positive initiatives on European construction.'[91]

Immediately after the EPP was established, preparations were put in place to found the EDU as an 'association of Christian Democratic and other non-collectivist parties'. The constitution was agreed on at a meeting in Munich in October 1977; December of the same year saw agreement on the issue of finance at a meeting in Vienna, and the EDU officially came into being at Schloss Klessheim, near Salzburg, on 24 April 1978.[92]

Once the EDU was in place, it became pointless for the EPP member parties to make any further attempt to agree on a joint strategy for organised dialogue with the Conservatives. Belgian, Dutch, Italian and French members of the EPP felt basic mutual trust had been put in question, and protested. There was a 'time-consuming and heated discussion about the EDU' at the May 1978 meeting of the EPP Political Bureau in Dublin, where the final version of the Political Programme was supposed to be decided on the basis of the EPP congress debates and resolutions. It was true that the founding of the EDU, within a few weeks of the EPP's successful first Congress, had an extraordinarily 'negative, divisive effect, tarnishing the image of the EPP'.[93]

'Simultaneous membership of the EPP and the EDU is bigamy', was the view of the Belgian PSC leader, Charles-Ferdinand Nothomb.[94] Henning Wegener, head of the CDU department of foreign relations from 1977 onwards, quoted Nothomb back to his party leadership. 'The formation of the EDU has caused serious ill-feeling in the EPP,' he said. 'Patient, painstaking explanation will be required to dispel it.' Addressing EPP bodies, he

---

[91] Cited in Gebauer, *Die europäischen Parteien der Mitte*, 154.

[92] On the founding of the EDU and its philosophy of Christian Democracy and Conservative community, see especially Franz Horner, *Konservative und christdemokratische Parteien in Europa: Geschichte, Programmatik, Strukturen* (Vienna and Munich, 1981), 79, and specifically on the founding and aims of the EDU, see Andreas Khol, 'Europäische Demokratische Union (EDU): Die europäischen Parteiengruppe der fortschrittlichen Mitte', in *Österreichische Monatshefte: Zeitschrift für Politik*, 5 (1978), 4ff.

[93] Joint meeting of the EPP and EUCD Political Bureaus, 6 June 1978, Berlin, Procès verbal: 'la réunion de Salzburg a eu un effet negatif de division, entamant l'image du PPE', ACDP IX-004-095 [translated from the 1998 version of the book].

[94] ADCP IX-004-095 [translated from the 1998 version of the book].

defended his party's view that the CDU considered the creation of the EDU to be necessary

- to forge links with like-minded parties in European countries where, for historical reasons, the Christian Democratic movement had not taken root;
- to pave the way for solid, non-collectivist majorities in the future European Parliament; and
- to offer a home to sister parties from European countries that did not yet belong to the European Community.[95]

The advent of the EPP as a specifically Christian Democratic European party with supranational ambitions led to a crisis within the EUCD, from which the EPP had sprung. It also led to the founding of the EDU, which was intended as both an alternative and a complement to the EPP. Subsequently, the EPP, EUCD and EDU – as was to be expected given their common origin – behaved like communicating bodies. By their nature, international political organisations, like national ones, strive for autonomy, a distinct identity and dominance. They invariably find subordination to, or dependence on, rivals hard to bear. In practice, a balance of sorts was struck, based, generally, on respect for a division of responsibilities. This is not to say, however, that things went as efficiently and as effectively as they should have. Eliminating the anomaly of having three transnational organisations within the political spectrum of the centre-right would turn out to be a rather difficult and long-term challenge.

---

[95]   Ibid.

Chapter Three

Rapprochement Towards Conservative and Other Like-minded Parties

After its founding in 1976, the European People's Party (EPP) welcomed a number of new member parties in accordance with the enlargement of the then European Community (EC). Following the accession of Greece in 1981, New Democracy (Néa Dēmokratía, ND) joined the EPP Group in 1982 and became a full EPP member in 1983. Despite its lack of affiliation with Christian Democracy, the centre-right ND was a preferred partner. Its pro-European stance was uncontested and its membership did not change the balance of power within the EPP, so its accession took place without much ado. The same was true for the Portuguese Democratic and Social Centre (Centro Democrático e Social, CDSp) and the Spanish Democratic Popular Party (Partido Demócrata Popular, PDP) – later renamed Christian Democracy (Democracia Cristiana), along with the Basque National Party (Partido Nacionalista Vasco, PNV) and the Catalonian Democratic Union (Unió Democràtica de Catalunya, UDC). These parties entered the EPP Group and became full members after Spain and Portugal joined the EC in 1986.

The electoral performance of the Iberian parties during European elections, however, was rather poor, especially compared with the Spanish and Portuguese Socialist parties that had significantly strengthened the Socialist Group in the European Parliament (EP). In the elections that followed the 1986 accession, for instance, only the Catalonian UDC managed to win a seat. This would not have been a major problem if relative strength in the EP did not matter. The contrary was true, however, and the Parliament's importance was likely to increase, given the rise in legislative power and competences it would be granted within the frame-

49

work of the Single European Act and, above all, in the Treaty of Maastricht.[96]

If the EPP was to avoid the danger of being marginalised, it would have to look for allies outside the traditional Christian Democratic world. So it did, seeking partners that shared the basic principles of the EPP. In spite of their different cultural traditions, the Christian Democratic parties of continental Europe were the Conservatives' natural allies. Ultimately, they shared the same values and represented the same electoral constituency. The rapprochement towards Conservative and other like-minded parties did not happen without internal dissent, however. Moreover, 'the Conservative question' that had bedevilled the EPP since its founding was a constant source of controversy in the EPP's ranks. The issue was fundamental to the EPP, since it had to do not only with strategy and power but also with the party's identity and essential coherence. It is no surprise, therefore, that the incorporation of Conservative and other like-minded parties was a delicate process that took more than a decade. Along with the enlargement towards the East, it transformed the EPP from an exclusive federation of Western European Christian Democratic parties into a pan-European political family of the centre-right.[97]

Table 1. EPP Member Parties from Western Europe (1976–2010)

Country	Party	Membership
Austria	Österreichische Volkspartei (ÖVP)	1995–
Belgium	Christelijke Volkspartij (CVP)/Christen-Democratisch & Vlaams (CD&V)	1976–
	Parti social chrétien (PSC)/centre démocrate Humaniste (cdH)	1976–
Denmark	Det Konservative Folkeparti	1995–
	Kristeligt Folkeparti/Kristendemokraterne	1993–

---

[96] Since the treaties are referred to here in terms of their political implications and not as legal documents, the term Treaty of Maastricht is used instead of Treaty on European Union (TEU).

[97] See, for example, Wilfried Martens, *Europe: I Struggle, I Overcome* (Brussels, 2009), chapters 6–8; and Steven Van Hecke, 'A Decade of Seized Opportunities: Christian Democracy in the European Union', in S. Van Hecke and E. Gerard, (eds.), *Christian Democratic Parties in Europe since the End of the Cold War* (Leuven, 2004), 269–95.

Table 1. (continued)

Country	Party	Membership
Finland	Kansallinen Kokoomus (KK)	1995–
	Suomen Kristillinen Liitto/Kristillisdemokraatit	2001– [a]
France	Centre des démocrates sociaux (CDS)/Union pour la démocratie française (UDF)	1976–2004
	Rassemblement pour la République/Union pour un mouvement populaire (UMP)	2001–
Germany	Christlich Demokratische Union (CDU)	1976–
	Christlich-Soziale Union (CSU)	1976–
Greece	Néa Dēmokratía (ND)	1983–
Italy	Democrazia Cristiana (DC)	1976–1994
	Partito Populare Italiano (PPI)	1994–2004
	Cristiani Democratici Uniti (CDUi)	1994–2002
	Centro Cristiano Democratici (CCD)	1994–2002
	Südtiroler Volkspartei (SVP)	1993– [a]
	Rinnovamento Italiano (RI)	1998–2004
	Forza Italia (FI)	1999–2009
	Unione Democratici per l'Europa (Udeur)/UDEUR Populari/Populari per il Sud	2001–
	Unione di Centro (UdC)	2002–
	Il Popolo della Libertà (PdL)	2009–
Ireland	Fine Gael (FG)	1976–
Luxemburg	Chrëschtlech-Sozial Volekspartei (CSV)	1976–
Netherlands	Christen Democratisch Appèl (CDA)	1976–
Norway	Høyre	1995– [b]
Portugal	Centro Democrático e Social (CDSp)	1986–1993
	Partido Social Democrata (PSD)	1996–
	Centro Democrático e Social-Partido Popular (CDS-PP)	2009–
San Marino	Partito Democratico Cristiano Sammarinese (PDCS)	1993– [a]
Spain	Partido Democrata Popular (PDP)/Democracia Cristiana	1986–1991
	Unió Democràtica de Catalunya (UDC)	1986–
	Partido Nacionalista Vasco (PNV)	1986–1999
	Partido Popular (PP)	1991–

Table 1. (continued)

Country	Party	Membership
Sweden	Moderata Samlingspartiet or Moderaterna (MS)	1995–
	Kristdemokratiska Samhällspartiet (KDS)/Kristdemo-kraterna (KD)	1995–

[a] observer status
[b] associate member

The Partido Popular

Not until the end of the 1980s did opening up the EPP to the Conservatives and other like-minded parties become an urgent item on the agenda. The issue was the situation in Spain. While only one of its three EPP member parties, the UDC, was represented in the EP, Members of the European Parliament (MEPs) from the conservative Popular Alliance (Alianza Popular, AP) became members of the European Democratic Group (EDG), where they joined the British and Danish Conservatives. The leader of the AP, Manuel Fraga Iribarne, was himself an MEP, and his group soon began to seek an alliance with the EPP Group. The Spaniards did not feel particularly at home in an EDG dominated by Britain's Tories, and it quickly became clear that they were far less conservative than their British EDG partners. This did not apply only to their policy on Europe, but also to their views on social policy, for instance. As a result of the close links they maintained with the Bavarian Christian Social Union (Christlich-Soziale Union, CSU), in particular, they increasingly acquired a taste for the EPP.

The Secretary General of the Council of Europe, Marcelino Oreja, although an independent, was known for his Christian Democratic views. It was under Oreja's influence that Fraga agreed, in the run-up to the 1989 European elections, to reform his party and thus enable political figures with a centrist profile, such as Oreja and his friends, to play a part in it.[98] Fraga had repeatedly been beaten in elections by the Socialists. He had

---

[98]   See Stefan Jost, *Die Politische Mitte Spaniens: Von der Union de Centro zum Partido Popular*, Saarbrücker Politikwissenschaft, vol. 18 (Frankfurt am Main, 1994), 255ff.

obviously become convinced that there was no chance of winning elections in Spain on a clearly right-wing ticket. He agreed to a change in his party's profile and programme, to its acceptance of the political forces of the centre and to its 'Europeanisation' based on the model of the EPP, which his party, thus reformed, would join. The credibility of this operation was to be sealed by his resigning from the party leadership and withdrawing from national politics.

In these circumstances, and in view of the hopelessness of the independent existence of the Democracia Cristiana, its Chair, Javier Rupérez, decided that his party should be dissolved and integrated into the successor of the AP, a process which was intended to give the party a completely fresh start. In this, Rupérez was vigorously supported by the EPP. At a meeting in Luxembourg in January 1989 with the party leadership (represented by the President Jacques Santer, the Group Chair Egon Klepsch and the Secretary General Thomas Jansen), Oreja and Rupérez agreed that the parties should act together and assist each other.

The right-wing Conservative AP was now transformed into a party of the centre-right: it was renamed People's Party (Partido Popular, PP) in the spring of 1989. And the Spanish Christian Democrats were prepared to join the PP. Everything was in place for integrating the PP into the EPP. The PP went into the 1989 European elections with a manifesto based on the EPP programme. Its MEPs – including some Christian Democrats like Oreja, who headed the list, and José Maria Gil-Robles – were admitted to the EPP Group after the elections. This was in line with a decision by EPP party leaders before the elections that any MEP who had stood for election on a list shared by Christian Democrats should automatically be allowed to join the EPP Group.[99] Although the electoral result was disappointing, the leadership did not question that both the EPP and the PP were heading in the right direction.

Soon after being elected to succeed Fraga as PP leader in April 1990, José Maria Aznar made contact with the EPP leadership to discuss the necessary steps for sustaining the cooperation, in other words, for joining the EPP. For his part, Aznar would have to overcome internal, right-wing conservative opposition to enter the EPP. Within the EPP, the resistance of the Catalans and Basques in particular was another obstacle: for internal

---

[99] Communiqué, Conference of Party and Government Leaders in Luxembourg, May 1989, EPP General Secretariat Archive (Brussels).

reasons, they opposed entry by the PP. The Italians, Belgians and Dutch, among others, had to be convinced that the Spanish party had undergone a genuine conversion and that the conditions existed for cooperation in a spirit of mutual trust and for a common policy at European level based on Christian Democratic principles.

In the meantime, Wilfried Martens, then still Prime Minister of Belgium, had been elected President of the EPP in May 1990. From the outset he had been a strong advocate of opening up the EPP and actively supported the PP's admission. The attitude of his own party, the Flemish Christian People's Party (Christelijke Volkspartij, CVP), was guarded to negative. In autumn of the same year, the PP was granted observer status. A year later, it became a full member of the EPP. As with the Greek ND, it soon became clear that the integration of the Spanish centre-right did not have to affect the EPP's Christian Democratic orientation or identity. Above all, the strong Spanish representation strengthened the EPP and its Group in pursuing its own policy goals in the EP.[100]

How to Open Up a Party

The opposition to admitting the PP proved that a more fundamental political line was needed to persuade those who were apprehensive about opening up the party to the Conservatives. More precisely, given the differing self-images and perceptions of member parties, there had to be a framework of principles which would ultimately convince, or at least reconcile, everyone concerned. Because the relationship between Christian Democracy and Conservatism was a sensitive and, historically speaking, highly charged issue, the aim had to be to prevent irreparable breaches within the EPP or between its member parties. This exercise was considered necessary, if not to say urgent, since the PP would not be the last non-Christian Democratic party to join the EPP – quite the contrary. The EC was destined to grow to include the Northern European countries in the relatively short term and the Central and Eastern European countries in the longer

---

[100] See also Steven Van Hecke, 'Europeanization and Political Parties: The *Partido Popular* and its Transnational Relations with the European People's Party', *International Journal of Iberian Studies* 22 (2009), 109–124.

term. The EPP, being a European party, naturally had to have a presence in all Member States if it was to remain credible and capable of making its mark. That also held true within countries where, for historical and cultural reasons, Christian Democratic parties had failed to emerge or had not developed into people's parties.

Further important reasons for pursuing a strategy of openness became apparent. The EPP is a party that has to fight for majority representation if it is to translate its ideas into reality. The EPP is a people's party and must not shut itself off from individual groups, classes or movements, least of all when any of these see themselves as having something in common with the party or want it to represent them. At the same time, the EPP is a Christian Democratic party and believes its ideas to be persuasive. The EPP is also a federalist party that respects the peculiarities of its regional or national components while seeking consensus between their different contributions in order to articulate and pursue a common policy.

To ensure that the EPP did not miss the chance of integrating allies or other like-minded forces, and to increase the party's influence across Europe so it could assume a leading role, the EPP had to see itself as – and make sure it was seen to be – an open people's party with Christian Democratic traditions and goals. A clear understanding had to exist that it was possible to work with those who represented other traditions and points of view, as long as they, in turn, respected the party's traditions and accepted its policies. Seizing this opportunity was also congruent with the EPP's ambition to become pre-eminent and, as far as possible, to translate its project into reality.

The party leadership took the decision to renew the EPP's basic 1976 programme and adapt it to the new circumstances and challenges the party faced. In President Martens's words: '[T]he expansion of the party would only prove durable and fruitful if there was agreement about the party's political foundations. Moreover, the acceptance of these fundamental principles had to be a basic condition for membership for new political parties. It was a fact that the greater the difference in parties the more important the common basis became.'[101] This process of fundamental renewal ultimately led to the adoption of a new basic programme at the 1992 EPP Congress in Athens, commonly known as the Athens Programme. At the

---

[101] Martens, *Europe: I Struggle, I Overcome*, 123.

level of its principles and basic ideas, the party was now prepared to incorporate Conservative and other similar parties.

At the level of political initiatives and activities, however, and amidst the process of EPP Group and party membership applications, the Conference of Party and Government Leaders took the decision at a meeting in Brussels on 13 April 1991 that the EPP 'would in future enter into closer cooperation with people's parties working towards a comparable social project in their countries, and towards the same goals as the EPP in European politics. As Europe's leading political force, the EPP is in principle prepared to admit those parties seeking membership provided that they accept the EPP's guidelines, basic policies and constitution'.[102] The decision to further open up the party was soon to be executed. The major issue was how to integrate British and Danish Conservative MEPs who had left the EDG after the Spanish centre-right had chosen the EPP.

British and Danish Conservatives

The British Conservative MEPs understood, before their party friends in Westminster or in the Tories' central office did, that their country's future was 'at the heart of Europe'. It was they who committed themselves to a rapprochement with the EPP. Directly after the 1989 European elections, the EDG, led by its Chair, Sir Christopher Prout, and with the agreement of its Danish members, applied for membership in the EPP Group in the EP. The application was so controversial that the Group felt unable to discuss it, let alone to take a decision, so EPP Group Chair Klepsch called together the EPP party executive. This body decided, after a lengthy debate in July 1989, that the time was not yet ripe, largely because of British Prime Minister Margaret Thatcher's policy on Europe. But Santer, the then party Chair, urged the Conservative and Christian Democratic Groups in the EP to intensify both dialogue and cooperation with each other. A fresh decision would be taken two years later in light of mutual experience.

---

[102] Procès-verbal, conférence des chefs de gouvernements et de partis, Bruxelles, 13 avril 1991, EPP General Secretariat Archive (Brussels) [author's translation].

Internal opposition in the EPP Group initially prevented the dialogue called for by the executive, or any closer cooperation, but in the summer of 1990 Martens began trying to fill the gap. He met more frequently with leading figures of the EDG, especially its Chair, Prout, and its Secretary General, Harald Rømer, and later also with the Chair of the British Conservative Party, Chris Patten. Urged on by Helmut Kohl, and under Martens's leadership, the April 1991 conference of EPP party leaders and heads of government once more considered the Conservatives' application. The conference mandated the EPP to establish by the end of the year whether, and under what conditions, a joint parliamentary group would be feasible, and under what circumstances.[103]

It was not until the second half of 1991 that dialogue between the EPP Group and the EDG began. Consultations were held on a number of sensitive issues.[104] The conclusion was that close cooperation was, in principle, possible. In February 1992, the conference of EPP party leaders and heads of government was able to report that given the outcome of these discussions, there was little to prevent the Christian Democrats and Conservatives from forming a single group. The EPP Group in the EP was given the mandate to set out the arguments for a joint parliamentary group with the Conservatives.[105]

Key preconditions for admitting the Conservative MEPs to the EPP Group, which eventually took place on 1 May 1992, were Thatcher's resignation as British Prime Minister and leader of the Tories and her successor John Major's willingness to pursue a different policy on Europe. But Kohl's determination was critical. Given the imminent establishment of the European Union (EU) and its enlargement, he had strategic reasons for wanting to open the EPP to the Conservatives, especially the British. Without Kohl, the breakthrough would not have come so soon. The decision could not have been taken at all, however, if EPP Chair Martens had not been so committed to the issue and had not taken political and practical responsibility for seeing it through.

After the British and Danish Conservative MEPs had joined the EPP Group, it was to be expected that in the medium term their parties would

---

[103] Ibid.

[104] Conclusions of the EPP/EDU working group (January 1992), EPP Group Archive (Brussels).

[105] Procès-verbal, conférence des chefs de gouvernements et de partis, Bruxelles, 14 février 1992, EPP General Secretariat Archive (Brussels).

also forge closer links with the EPP. The first step was taken by the Danes when, in July 1993, the Conservative People's Party (Konservative Folkeparti) applied for permanent observer status. In 1995, the party became a full EPP member, together with the Danish Christian People's Party (Kristeligt Folkeparti, since 2003 Kristendemokraterna). The British Conservatives, however, never applied for party membership, although the prospect had been relatively feasible, especially with leading figures like Prout and Patten.

After Major's resignation as Prime Minister and Tory Party leader in 1997, the Tories' relationship with the EPP and its Group went up and down, but many times more down than up.[106] The dominance of Eurosceptics in the British delegation after the 1994 European elections proved a bad omen. With party leader William Hague, who succeeded Major in 1997, relations with the EPP started to deteriorate seriously. In 1999, however, the British Conservatives won the European elections. Their MEPs were necessary for the EPP Group to become, for the first time since direct elections in 1979, the largest group in the EP. A deal was made in July 1999 under which British Conservatives continued to sit with EPP deputies, while the Group name was changed to Group of the European People's Party (Christian Democrats) and European Democrats. The addition of European Democrats (ED) made it clear to everyone, not least the Tories back home, that the British Conservatives did not any longer belong to the 'pro-European' EPP Group.

In 2001, Hague was replaced as party leader by Ian Duncan Smith. Not a fan of further European integration, either, on several occasions he began to threaten to leave the EPP-ED Group. A true Euro-sceptic – he voted against the Maastricht Treaty – Smith asked for 'a special status', giving more freedom and privileges to the British Conservatives in the EP. The issue was regularly debated at EPP Summits, sometimes in the presence of Smith. His successor, Michael Howard, party leader between 2003 and 2005, eventually made an agreement with then Group leader Hans-Gert Pöttering to stay with the Group. The concession made by the Group, for which its Internal Regulations had to be changed, was that the British Conservatives were granted the right to develop and to promote deviant opinions with regard to institutional issues, such as the European Constitution. Proponents of this agreement claimed that it only made official a situation

---

[106] Martens, *Europe: I Struggle, I Overcome*, 158–60.

that had long existed. Opponents considered this concession a bridge too far. What, for instance, if other parties or delegations asked for the same consideration or simply joined the British Conservatives? Nonetheless, leading up to the 2004 European elections a majority of the deputies voted in favour of the amendments to the Internal Regulations which made it possible for the British Conservatives to continue to be 'ED' within the EPP-ED Group.

A final phase, at least so far, started in 2005, when David Cameron was elected leader of the Tories. During his campaign as party leader he promised to leave the EPP-ED Group and form his own group as soon as possible. This was clearly done for internal party reasons, namely to secure the votes of his Euro-sceptic party members. Most of the British Conservative MEPs had always been in favour of collaboration with the EPP deputies. In the course of 2006 and 2007, he set up the so-called Movement for European Reform, along with the Czech Civic Democratic Party (Občanská demokratická strana, ODS). Eventually it proved to be a vehicle for the preparation of a new group in the EP. In the run-up to the 2009 election it was already clear that the British Conservatives would not continue to sit in the ED section of the EPP-ED Group. With the ODS and a number of other parties, they created a new EP group, the European Conservatives and Reformists (ECR). This heterogeneous group did not prevent the 'new' EPP Group from remaining the largest body in the EP after the elections. The EPP Group's cohesiveness grew, and new opportunities to work more closely with the party emerged. The EPP Group, however, lost its representation in the UK, one of the largest and most important EU Member States.

The Nordic Conservatives

The EPP's rapprochement towards the Nordic Conservatives was of a different kind. It started in the period after the end of the Cold War and picked up speed in the framework of the 1995 enlargement of the EU. The process differed from that for the countries discussed so far, since in many of the Nordic countries Conservative parties existed in addition, and sometimes in opposition, to Christian Democratic parties – parties that were already members of the EUCD. As early as 1989, for instance, the Finnish Conservative party, the National Coalition Party (Kansallinen Kokoomus,

KOK) had expressed an interest in becoming involved in the EUCD and the EPP. In spring 1991, the KOK applied for observer status in the EUCD, but was rejected by the Political Bureau because of opposition from representatives of various member parties, notably the Scandinavian Christian Democratic parties.

Scandinavia's and Finland's Christian Democratic parties are much younger than their equivalents in continental Europe. They were founded in reaction to the secularisation of public life in the 1960s and 1970s, a process promoted not only by the Socialists and Liberals but also by the Conservatives. Apart from the Norwegian Christian People's Party (Kristelig Folkeparti), founded as early as the 1930s, none of them existed before the 1960s. Rather than Christian Democratic parties, they should be seen as Christian confessional parties, taking a stand for the teachings of their churches in politics. Apart from the Swedish Christian Democratic Coalition Party (Kristdemokratiska Samhällspartiet, KDS), these parties have so far found it difficult to formulate a positive approach to the process of European integration.[107] In the countries of mainland Europe, middle-class values and people's parties are normally Christian Democratic. In Scandinavia, this representation has tended to fall to Conservative parties. However, it is also clear that the Christian parties of Norway and Sweden have potential for development and growth. This is expressed in the way they have been able to hold their own deputies in their respective parliaments and occasionally take over the reins of government. As a result they have been exposed to influences which in mainland Europe have turned confessional parties into people's parties.

Immediately after Carl Bildt, the Chair of Sweden's Moderate Coalition Party (Moderata Samlingspartiet or Moderaterna, MS), formed a coalition government with the Liberals, Centrists and Christian Democrats in the autumn of 1991, the party also expressed an interest in joining the EPP. The EPP leadership now pressed hard for the necessary dialogue with the northern Conservatives to be accompanied by a special effort to reach agreement with Christian Democrats in these countries. This was essential if minds inside the EPP were to be changed. Any involvement of the Conservatives in the Christian Democrats' European political organisation was bound to be controversial, since at home in Scandinavia they were rivals.

---

[107] In 1996 the KDS changed its name to Christian Democrats (Kristdemokraterna, KD).

For a number of EPP member parties, the way in which their future partners in Scandinavia dealt with this problem would determine whether or not they agreed to extend EPP membership to them.

The first step was the admission of the Swedish KDS. A provision in the EPP's constitution allowed EUCD member parties in countries wanting to join the EC to be admitted to the EPP as associate members, and referring to that, the KDS submitted an application. It was approved in the autumn of 1991. This step was initially controversial. Unaware of the political situation in Sweden, the representatives of some member parties in the EPP Bureau assumed that the KDS was insignificant, *quantité négligeable*. They feared that any recognition of this small, relatively young party – without a single MP in the Swedish Parliament until the September 1991 elections – would hold up recognition of the MS. An additional factor was that anti-European statements by the Christian parties of Norway, Denmark and Finland were mistakenly attributed to the KDS.

In view of their weak position, the Scandinavian Christian Democratic parties' concern about the Conservative parties drawing closer to European Christian Democracy was understandable. The Swedish KDS, in particular, responded in a constructive way to the sympathy expressed by the EPP leadership on this matter. The KDS leadership understood and accepted that the European view of things differed from their national and Scandinavian view, and that the EPP's essential concept of Christian Democracy went beyond the positions underlying the self-image and programmes of individual parties. It must also be borne in mind that the conservatism of Sweden's MS, Finland's KOK, Norway's Høyre and Denmark's Konservative Folkeparti is not the same as that of the British Conservatives, let alone of Thatcherism. These are parties of the centre that are open to Europe and advocate a social market economy. Their position is moderate and middle-class. This, too, was a factor in smoothing the process of convergence and the incorporation of the Nordic Conservatives into the EPP.

To enable the Scandinavian parties to work with it, the EPP had to amend its statutes. As noted above, associate membership or observer status could be granted to parties from countries seeking to join the EC, but only when the applicant parties were already members of the EUCD. The Athens Congress passed the required change in the statutes in November 1992. Two months later, in January 1993, MS and KOK were accepted as permanent observers. This decision had near-unanimous approval, once the KDS representatives signalled their agreement. In May 1993, the EPP's Political Bureau also accepted Norwegian Høyre's application with no votes against it, nor even any abstentions. After Sweden's and Finland's accession to the EU, MS and KOK were admitted as full members in February 1995, together with the Danish Conservatives. As Norway did not join the EU, Høyre could not become a full EPP member but was granted the status of associate member.[108] In 2001, the Finnish Christian Democrats (Kristillisdemokraatit) were granted observer status.[109]

The Austrian People's Party

Along with Sweden and Finland, Austria also became a member of the EU in 1995. Unlike many of the parties applying for close cooperation with the EPP and its Group in the EP, the Austrian People's Party (Österreichische Volkspartei, ÖVP) was well known to the EPP. This party was part of the continental Christian Democratic tradition, had much in common with centre-right parties in Western Europe and was a forerunner in terms of party cooperation at the European level. Due to Austria's neutral position during the Cold War and its consequent position outside the European integration process, the ÖVP was unable to become a full member of the EPP in 1976, much to its regret. Instead the ÖVP became one of the driving forces, perhaps the strongest, in the EDU. This did not prevent the ÖVP from applying for EPP membership as soon as possible after the end of the Cold War. Soon after Austria's accession to the EU, the ÖVP became a full EPP member.

The Portuguese Social Democrats

With the arrival of the Portuguese Social Democratic Party (Partido Social Democrata, PSD) in the autumn of 1996, the EPP gained a new member party, one which since its founding in the 1970s had belonged to the Liberal International and the European Liberal Democrat and Reform

---

[108] Decision of the EPP executive in February 1995, EPP General Secretariat Archive (Brussels).

[109] Until 2001 the party was called Finnish Christian League (Suomen Kristillinen Liitto).

Party (ELDR). Despite its name, which evidently reflected the political realities at the time it was founded, the PSD, under the direction of Anibal Cavaco Silva, had become an authentic popular party since the mid-1980s, able to mobilise absolute majorities in the political centre. Those in the circle of the CDSp, such as Diogo Freitas do Amaral and Francisco Lucas Pires, certainly played a part in this development. When, in 1991, the CDSp was thrown out of the EPP because of its drift towards Euro-scepticism, the pro-Europeans found a new political home in the PSD. With the PSD deputies on board, the EPP Group once more had a strong Portuguese delegation. And with José Manuel Durão Barroso, President of the PSD between 1999 and 2004, the EPP could also boast a prime minister from Portugal among its ranks between 2002 and 2004. In 2009, the rightist Democratic and Social Centre-People's Party (Centro Democrático e Social-Partido Popular, CDS-PP), the successor of the CDSp, also became a full member of the EPP.

Forza Italia

The collapse of Italy's Democrazia Cristiana (DC) in the early 1990s was a critical development. This party had been the EPP's third essential pillar, alongside the German and Benelux Christian Democratic parties. The new Italian party, Forward Italy (Forza Italia, FI), now advanced itself as the potential new partner from Italy, along with DC's small successor parties, the Italian People's Party (Partito Popolare Italiano, PPI), the Christian Democratic Centre (Centro Cristiano Democratico, CCD) and the United Christian Democrats (Cristiani Democratici Uniti, CDUi), which were all recognised as full EPP member parties. FI, Silvio Berlusconi's party, had won by far the largest of the Christian Democratic constituencies. However, for reasons connected with Italian as much as European politics, no link was at first forged between the EPP and FI.

After the 1994 European elections, FI deputies created their own single-nationality group, named Forward Europe (Forza Europa). In early July 1995, this was fused with the parliamentary group European Democratic Alliance (EDA), made up of deputies from, among others, the French Gaullist party Rally for the Republic (Rassemblement pour la République, RPR) and the Irish Republican Party (Fianna Fáil). Together they created the Union for Europe Group in the EP. Berlusconi was still keeping an eye on the EPP, however, and in 1997, after much delay, he finally managed to establish a so-called contact group between FI and the EPP Group. Opposition to a rapprochement with FI remained widespread within the EPP, but when Berlusconi announced at the end of 1998 the creation of a new European party called Union for Europe, similar to the Group in the EP, things started to change. The EPP leadership was convinced this new development would threaten the position of the EPP as the main representative of the European centre-right, and also generate divisions within some member parties about which alliance would be most beneficial for them to join. To avoid this scenario, contacts with FI were quickly established, both at the level of the EP and at the level of party presidents. Early in 1998, at an informal meeting of EPP government leaders and the EPP leadership, it was decided that everything should be done to enlarge the EPP and incorporate, among other parties, FI.[110]

In June 1998, FI MEPs joined the EPP Group. One and a half years later, FI was finally accepted as a full member of the EPP. In the meantime, Italian Renewal (Rinnovamento Italiano, RI) had also joined the EPP. Founded by Lamberto Dini, a former Prime Minister of Italy and President of the Italian Central Bank, this was another of the new parties to appear after the implosion of the Italian party system in the early 1990s. Dini's party became the new home of Liberal Democrats, who had for the most part been members of the DC. Counting the three direct successor parties of the DC, there were now five Italian parties that were member parties of the EPP, plus the South Tyrolean People's Party (Südtiroler Volkspartei, SVP), which had had observer status since 1996.

The EPP had always supported the joining of forces of the Italian centre-right. Those parties that directly succeeded from the DC had always opposed this idea. Several attempts were made, none of them successful. In 2002, for instance, CCD, the party of Pier Ferdinando Casini, was succeeded by Unione di Centro (UdC), an umbrella party of several small Christian Democratic parties, including the CDUi, led by Rocco Buttiglione. The party officially became a full EPP member in the same year. Also in 2002, RI, Dini's party, joined the centre-left The Daisy (La Margherita) led by Francesco Rutelli, which later merged into the Democratic Party (Partito Democratico), an umbrella party led by former Prime Minister and European Commission President Romano Prodi. The Union of

---

[110] See Martens, *Europe: I Struggle, I Overcome*, 142–4.

Democrats for Europe (Unione Democratici per l'Europa, Udeur), the party of Clemente Mastella, became an EPP member in 2001. The party transformed into UDEUR Populars (UDEUR Populari) and in June 2010 was replaced by the Populars for the South (Populari per il Sud). The largest among the Christian Democratic parties, the PPI, also joined the Democratic Party in 2007. Prior to the 2004 European elections, the PPI left the EPP and the EPP-ED Group in the EP.

Eventually a regrouping of the centre-right would occur around FI. After an initial attempt was launched in 2007, and a lot of parties showed interest, in March 2009 The People of Freedom (Il Popolo della Libertà, PdL) was established.[111] The new party, which was created in the run-up to the 2009 European elections, was mainly concentrated around Berlusconi's FI and Gianfranco Fini's National Alliance (Alleanza Nazionale, AN). They performed very well in the European elections and their delegation became one of the cornerstones of the EPP Group. Immediately after the elections, the EPP granted full membership to this new party.

The French Gaullists

Along with the deepening of the European integration process with, among other things, more power and competences granted to the EP and enlargement of the Union's membership, party decline also played a major role in the EPP's representation in France. The French Christian Democrats of the Democratic and Social Centre (Centre des démocrates sociaux, CDS), who had been co-founders of the EPP, had been in gradual decline over the course of the 1970s. The idea that they could once more become the pivotal party in the French political system faded away. Instead, French Christian Democrats were constantly looking for allies, at the national as well as the European level. In 1978 the CDS became part of the Union for French Democracy (Union pour la démocratie française, UDF), together with the Liberals and the Republicans.

Approaching the 1994 European elections, the French centrists of the UDF joined forces with the Gaullists. Dominique Baudis, a Christian Democrat and Mayor of Toulouse, where the 1997 EPP Congress was to take place, headed the list. Jacques Chirac, President of the RPR, and

---

[111] In 2008 a joint list called PdL won the general election.

Valéry Giscard d'Estaing, the UDF President, reached agreement that all European deputies elected on the joint list of the RPR and UDF governing parties should join the EPP Group. After the elections, however, the RPR leadership did not honour the deal with the EPP, and instead renewed their alliance with the EDA.

When FI left this alliance, and because the EPP was actively seeking to establish a closer relationship with it, the RPR also chose the EPP. After the 1999 European elections, RPR deputies joined the EPP-ED Group in the EP. Two years later, in December 2001, the RPR joined the EPP as a full member. The level of consensus about bringing in the RPR was emblematic of the degree to which even the leading critics of the inclusion of Conservative parties had changed their minds. But of course the RPR had changed radically too. Parliamentary cooperation had caused neither ideological nor practical difficulties. The RPR's radical Gaullist, nationalistic wing, led by Charles Pasqua, had split during the internal party debates about ratifying the Maastricht Treaty, and when the party leadership passed from Philippe Séguin to Nicolas Sarkozy (1999) and then to Michèle Alliot-Marie (1999–2002), the party found a new equilibrium. Ahead of the French presidential election in 2002, the political ground shifted. An electoral alliance led by Alain Juppé and called the Union for a Presidential Majority (Union pour une majorité présidentielle) became in April 2002, following its electoral victory, a party, namely the Union for a Popular Movement (Union pour un mouvement populaire, UMP). Many Christian Democrats, as well as other politicians from the UDF, joined the new French people's party, which as a successor to the RPR remained a full member of the EPP.

Having at last a firm representation in France, the EPP now had strongholds in all Member States of the EU except for the United Kingdom, where until 2009, the British Conservatives only collaborated with the EPP in the EP. Equally important, because the party's enlargement to include Conservative and other like-minded parties was accompanied by a process of deepening, resulting in the 1992 Athens Programme, the debate on opening up the EPP had also made it politically more cohesive and more serious. In other words, the EPP was now ready to turn to the East, another major challenge, but one that had been prepared for well in advance.

Chapter Four

Eastward Enlargement

From the early 1990s onwards, the issues raised by the admission of new members were central to the internal development of the European People's Party (EPP). There were, for instance, the changes in the Italian party system and within the French centre-right, as well as the 1995 enlargement of the European Union (EU). Probably the biggest challenge was the prospect of further enlargement to 20 or even more Member States, most of them former Communist countries from Central and Eastern Europe. This, in turn, led to plans to include the Western Balkans. It also triggered an eastward enlargement within the EPP that included new full members, associate members and parties with observer status.

The EPP, however, was not the only organisation that played a role in this process of party enlargement. The European Union of Christian Democrats (EUCD), institutionally close to the EPP and entangled with its history, was also involved. The European Democratic Union (EDU), in which several major EPP parties worked together with Conservative parties, also played a part in ensuring that potential allies or member parties in practically every one of the future EU states were brought into the political family. Together these organisations permitted a certain oversight in the ever-shifting pattern of parties in Central and Eastern Europe. At the same time, it reopened the question of whether there should be three separate centre-right organisations in Europe.[112]

---

[112] On the founding of the EUCD and the EDU, see chapters one and two, respectively, of the present volume. On the road to the merger of the EPP, EUCD and EDU, see chapter five.

The EUCD and the Robert Schuman Institute

Before parties from Central and Eastern Europe became officially affiliated with the EPP, the bulk of the preparatory work was done by the EUCD. In the years after the fall of the Berlin Wall in 1989, the Italian Christian Democrat Emilio Colombo chaired the EUCD. In January 1993 he was succeeded by EPP President Wilfried Martens. In this way, the leadership of the EPP and the EUCD were de facto unified through one person. Under Martens's leadership an important decision was made. In 1995 'the Union of the Robert Schuman Institute for Developing Democracy in Central and Eastern Europe' was founded by the EUCD, the Robert Schuman Foundation, the Foundation for European Studies, the EPP, and a number of national parties and other organisations. The Robert Schuman Institute replaced the Christian Democratic Academy for Central and Eastern Europe that had been founded in 1991.[113]

The Robert Schuman Foundation had been established in 1989 by a number of leading Christian Democrats, most of them Members of the European Parliament (MEPs), along with others such as EPP Group Chair Egon Klepsch.[114] According to the Foundation's statutes, it was created on the basis of Christian ideals and Christian Democratic policies to 'promote the training and further training of gifted young people of suitable character, provide political education, take steps to strengthen democracy and pluralism in Europe and the world, encourage international cooperation, make the results of its activities available to the general public by the production of publications and support moves towards European union.'[115] As a Budapest-based Hungarian NGO created by the Robert Schuman Foundation, the EPP and other organisations, the Robert Schuman Institute concentrated on the countries of Central and Eastern Europe. As this area became the central focus of the EU's enlargement policy, the Robert Schuman Institute gained importance within the EUCD accordingly.

---

[113] This was decided by the EUCD and EPP in Strasbourg on 12 July 1991. See Erhard von der Bank and Kinga Szabó (eds.), *Robert Schuman Institute. Fifteen Years for Developing Democracy in Central and Eastern Europe* (Budapest, 2006), 10, 31.

[114] The EPP's Robert Schuman Foundation with its seat in Luxembourg should not be confused with the Paris-based Fondation Robert Schuman (www.robert-schuman.eu).

[115] Article 2 of the 1989 statutes of the Robert Schuman Foundation.

As part of a long and complex process of social and economic transformation in Central and Eastern Europe, the Robert Schuman Institute was founded to contribute to the stabilisation and development of democratic political systems, as well as the introduction of a market economy. It also directed its activities towards an even more ambitious goal: the changing of people's mindsets and the strengthening of civil society. Political education has always played a very important role in this respect. The assistance offered by the Robert Schuman Institute should be seen as a concrete application of the Institute's basic objectives to

- promote the idea of a United Europe;
- support and foster the process of democratic transformation in the CEE/EE/SEE [Central, Eastern, and South-Eastern European] countries on the basis of European Christian Democratic values in the spirit of Robert Schuman;
- promote the development of civil societies in the CEE/EE/SEE countries;
- facilitate the information flow and contacts between the European Union, the acceding countries and non-EU member states;
- contribute to the fundamental programme of the EPP member parties in the EU and their partner parties in the CEE/EE/SEE countries;
- assist in preparing the governing or opposition partner parties in the CEE/EE/SEE countries; and
- establish and operate a documentation centre to serve a more efficient exchange of information among the different organisations within the EPP family.[116]

By giving education and training to mostly young politicians of the member parties of the EPP, as well as by organising conferences on current problems of the region, the Robert Schuman Institute continues its work, even after the 2004 and 2007 enlargement of the EU. The focus is now on the Western Balkans and Eastern Europe. In 2010 Austrian MEP Othmar Karas, Vice-President and Treasurer of the EPP Group, became President of the Robert Schuman Institute. In 2002 the Institute was joined by the Centre for Political Parliamentary Education and Training (CET), founded by the then EPP-ED Group in the European Parliament (EP). The CET supported member parties and their parliamentarians. Although it

---

[116] See 'The Institute: Short History' at the Foundation's website, www.schuman -institute.eu.

paid particular attention to candidate countries, it also worked with MEPs and their assistants, parliamentary candidates and officials in the Group and in the Parliament. The CET functioned as an external office of the General Secretariat of the EPP Group in Budapest, similar to its offices in Berlin, London, Madrid, Paris and Rome. Its work was undertaken in close cooperation with the Robert Schuman Institute.[117]

Further Preparatory Work

Once more stable parties had been established, the EPP could start think- ing about welcoming allies from Central and Eastern Europe. For such a delicate operation, the EPP deployed a number of tools. A range of sub- organisations prepared the entry of new parties into the EPP. Apart from its own Robert Schuman Institute and the EUCD, the EPP made use of the expertise and reports of the Konrad Adenauer Foundation (which has of- fices in a number of Central and Eastern European countries) and the work done by its own Working Group 'Central and Eastern Europe', renamed 'Enlargement' and, most recently, 'EPP Membership'.[118] The following method for dealing with membership applications has been established: the Political Assembly (formerly the Political Bureau) examines the appli- cation and asks the Working Group for advice. The Working Group pre- pares an application report with recommendations. This report is often based on a fact-finding mission to the country in question and covers meet- ings with the applicant party and other political parties, as well as non- governmental organisations (NGOs); it also includes party programmes and manifestos. The report from the Working Group is then referred to the Political Assembly, which takes the final vote on the application.[119]

The EPP not only created a special Working Group to accommodate this challenge, it also changed its statute. At the 1997 Toulouse Congress a

---

[117] Von der Bank and Szabó, *Robert Schuman Institute*, 5, 12, 115. See the CET's website, www.eppgroup-cet.eu.

[118] The Working Group was for a long time chaired by Dutchman Wim van Vel- zen. In 2004 he was succeeded by Camiel Eurlings. Since 2007 the Working Group has been chaired by Corien Wortmann-Kool, also from the Dutch CDA.

[119] Since the EPP currently counts more than 70 member parties, around 60 mem- bership applications have succesfully been completed, not all, but the bulk of them coming from Central and Eastern Europe.

special membership category was introduced. In addition to observer status and full membership, associate membership was created for parties from countries that were not yet members of the EU. Their status as associate members meant that they could take part in all committees, activities and meetings, with more or less the same rights as full members. Formally, full membership was still restricted to parties from EU Member States, but within non-EU member states a distinction was made between parties from non-candidate countries (observer status) and parties from candidate countries (associate membership). These three stages are not automatic, however. Each step or membership upgrade has to be approved by the Political Assembly. Where a membership application is regarded as premature because of internal divisions, political instability or any other reason that raises questions within the EPP, the dossier is often reconsidered at a later stage.

Most membership applications from Central and Eastern Europe were considered in clusters, the first wave taking place in the second half of the 1990s. Sometimes applications from competing parties within one country were discussed together, sometimes not. Being granted observer status was often the most difficult hurdle for applicants. Given the unstable political situation in many former Communist countries, especially in the period immediately after the fall of the Berlin Wall, the EPP took the application procedure seriously. In most cases the way in which the EPP handled the applications proved to be successful, with new parties contributing to the EPP and moving on to become full members. In some cases, however, parties with observer status ceased to exist after a while, mostly due to severe electoral losses. In many countries the EPP pleaded actively for cooperation between like-minded parties, even, if possible, for an electoral cartel or merger. The high number of centre-right parties competing, sometimes primarily among themselves, was a clear disadvantage vis-à-vis the left, and also made collaboration with the EPP more difficult.[120] The easiest of the applications were from parties that had historical roots in nineteenth and early twentieth century Christian Democracy. They often had links with other member parties that dated back to the period before the Second World War, when they had been engaged in the international organisation

---

[120] Wilfried Martens, *Europe: I Struggle, I Overcome* (Brussels, 2009), 196: '. . . only understanding and cooperation between the members of the same political family can lead to a lasting result'.

of the Christian Democracy of that period.[121] But those parties were a minority, and over all, performed less well than might have been expected. As in Western Europe, therefore, the EPP also welcomed centre-right parties that were not Christian Democratic but were willing to subscribe to the party's basic values and ideas.

Geographically, Cyprus and Malta are not Central and Eastern European countries. Moreover, they belonged, politically speaking, to the West. The Democratic Rally of Cyprus (Dimokratikos Synagermos, DISY) was an EDU member, while the Maltese Nationalist Party (Partit Nazzjonalista, PN) belonged to the EUCD. They had already become associate members of the EPP in 1991 and 1994, respectively. Their long-term leaders, Glafkos Clerides and Eddie Fenech Adami, attended the EPP Summits regularly. Since 1997 DISY has been chaired by Nikos Anastasiades, while since 2004 the PN has been led by Lawrence Gonzi.

Table 2. EPP Member Parties from Central and Eastern Europe (1989–2011)[122]

Country	Party	Observer Status	Associate Member	Ordinary Member
Albania	Partia Demokratike e Shqipërisë (PDa)	2003–		
Belarus	Bielaruski narodny front (BPF)	2006–		
	Abjadnanaja hramadzianskaja partyja Bielarusi (UCP)	2006–		
Bosnia and Herzegovina	Stranka demokratske akcije (SDA)	2004–		
	Hrvatska demokratska zajednica Bosne i Hercegovine (HDZBiH)	2004–		
	Partija demokratskog progresa (PDPbih)	2004–		
Bulgaria	Graždani za Evropeĭsko Razvitie na Bǎlgarija (GERB)			2008–
	Sǎjuz na Demokratičnite Sili (UDFb)			2007–
	Zemedelski Naroden Sajuz (ZNS)			2007–
	Demokrati za Silna Bǎlgarija (DSB)			2007–
	Demokratičeska Partija (DP)			2007–

[121] See chapter one of the present volume, particularly the International Secretariat of Democratic Christian-inspired Parties (Sécretariat international des partis démocratiques d'inspiration chrétienne, SIPDIC).

[122] Data as of 1 March 2011.

Table 2. (continued)

Country	Party	Observer Status	Associate Member	Ordinary Member
Croatia	Hrvatska demokratska zajednica (HDZ)		2004–	
	Hrvatska seljačla stranka (HSS)		2007–	
	Demokratski centar (DCc)	2002–		
Cyprus	Dimokratikos Synagermos (DISY)			2004–
Czech Republic	Křesťanská a demokratická unie–Československá strana lidová (KDU-ČSL)			2004–
	Tradice, Odpovědnost, Prosperita 09 (TOP09)			2011–
Estonia	Erakond Isamaaliit–Pro Patria Union			2004–2006
	Erakond Res Publica			2004–2006
	Isamaa ja Res Publica Liit (IRL)[a]			2006–
FYROM	Vnatrešna Makdonska Revolucionerne Organizacija–Demokratska Partija za Makedonsko Nacionalno Edinstvo (VMRO-DPMNE)		2010–	
Georgia	Ertiani Natsionaluri Modzraoba (UNM)	2008–		
Hungary	Fidesz-Magyar Polgári Szövetség (FIDESZ)			2004–
	Kereszténydemokrata Néppárt (KDNP)			2007–
	Magyar Demokrata Fórum (MDF)			2004–2009
Latvia	Jaunais Laiks (JL)			2004–
	Tautas Partija (TP)			2004–
	Pilsoniskā Savienība (PS)			2010–
Lithuania	Tėvynės Sąjunga (TS)			2004–2008
	Lietuvos krikščionys demokratai (LKD)			2004–2008
	Tėvynės Sąjunga–Lietuvos krikščionys demokratai (TS-LKD)[b]			2008–

Table 2. (continued)

Country	Party	Observer Status	Associate Member	Ordinary Member
Malta	Partit Nazzjonalista (PN)			2004–
Moldova	Partidul Popular Creştin Democrat (PPCD)	2005–		
	Partidul Liberal Democrat din Moldova (PLDM)	2011–		
Poland	Platforma Obywatelska (PO)			2004–
	Polskie Stronnictwo Ludowe (PSL)			2004–
Romania	Partidul Democrat-Liberal (PD-L)			2007–
	Uniunea Democrată Maghiară din Ro-mania/ Románia Magyar Demokrata Szövetség (RMDSz)			2007–
	Partidul Naţional Ţărănesc Creştin Democrat (PNŢCD)			2007–
Serbia	G17 PLUS		2003–	
	Demokratska stranka Srbije (DSS)		2003–	
	Savez vojvođanskih Mađara–Vajdasági Magyar Szövetség (VMSZ)	2007–		
Slovakia	Kresťanskodemokratické hnutie (KDH)			2004–
	Slovenská demokratická a kresťanská únia–Demokratická strana (SDKÚ-DS)			2004–
	Strana maďarskej koalície–Magyar Koalíció Pártja (SMK-MKP)			2004–
Slovenia	Slovenska demokratska stranka (SDS)			2004–
	Nova Slovenija–Krščanska ljudska stranka (NSi)			2004–
	Slovenska ljudska stranka (SLS)			2004–
Turkey	Adalet ve Kalkınma Partisi (AKP)	2005–		
Ukraine	Batkivschyna	2008–		
	Narodnyi Soyuz Nasha Ukrayina (NSNU)	2005–		
	Narodnyi rukh Ukraïny (RUKH)	2005–		

[a] since 2006 merger of Erakond Isamaaliit–Pro Patria Union and Erakond Res Publica

[b] since 2008 merger of Tėvynės Sąjunga and Lietuvos krikščionys demokratai

The Entry of New Parties

Poland, the biggest country of Central Europe, was for a long time a difficult case for the EPP. The first Polish party to become an observer was the Freedom Union (Unia Wolności, UW). It had roots in the Solidarity trade union movement and was founded in 1994 as a merger between two parties. The Polish UW was prepared for its inclusion in the structures of transnational party cooperation by its membership in the EDU. The UW, however, lost all its seats in the Sejm, the lower house of the Polish Parliament, in the 2001 elections. After the dissolution of the EUCD, the following parties also joined the EPP: the Conservative People's Party (Stronnictwo Konserwatywno-Ludowe, SKL) and the electoral alliance that emerged from the Polish Solidarity movement, Ruch Spoleczny (RS-AWS). RS was the leading party within the Solidarity Election Action (Akcja Wyborcza Solidarność, AWS). Jerzy Buzek, later to become President of the EP, chaired as Prime Minister a centre-right AWS-UW government coalition between 1997 and 2001, followed by a short-lived AWS minority government. While the main leaders of the SKL decided to join the Civic Platform (Platforma Obywatelska, PO), the RS-AWS was dissolved before it could become a full EPP member. Meanwhile, two other parties had started to establish contacts with the EPP. PO was founded in 2001.[123] Donald Tusk became party leader in 2003. In the same year PO was granted associate membership. Following Poland's entry into the EU, PO became a full member in 2004. Also, the Polish People's Party (Polskie Stronnictwo Ludowe, PSL) became a full member in 2004. In 2007 it entered a government coalition with PO, led by Tusk.

In Czechoslovakia, two separate party systems, Czech and Slovak, developed in the two years between the fall of the Communist regime at the end of 1989 and the split of the country into the Czech and Slovak republics in January 1993. In the Czech Republic, the Czechoslovak People's Party (Československá strana lidová, ČSL) dated back to the founding of the Republic , having been established in 1919. It continued to exist as a 'block party' (Blockpartei) during Communist rule. After the Velvet Revolution of 1989, the party was renamed the Christian and Democratic

---

[123] In the same year the Kaczyński brothers founded the Law and Justice Party (Prawo i Sprawiedliwość, PiS). At first they were also in favour of membership in the EPP, and their MEPs became members of the EPP-ED Group.

Union–Czechoslovak People's Party (Křesťanská a demokratická unie–Československá strana lidová, KDU-ČSL). As early as 1990 the party became a full member of the EUCD. In 1990 the Civic Democratic Alliance (Občanská demokratická aliance, ODA) also became an EPP observer. The party, founded in 1989, ceased to exist in 2007. It took part in the first government of the independent Czech Republic, headed by Václav Klaus. This government was composed of Klaus's Civic Democratic Party (Občanská demokratická strana, ODS), KDU-ČSL, ODA and the small Christian Democratic Party (KDScz). The MEPs of the ODS composed, together with the British Conservatives, the 'ED' in the EPP-ED Group between 2004 and 2009. The ODS never formally applied for EPP membership, although its leader between 2002 and 2010, Mirek Topolánek, showed interest. He even participated in some EPP Summits.[124] Nevertheless, KDU-ČSL became an observer in 1997 and a full member in 2004. ODS's splinter party, the Freedom Union (Unie svobody) became an associate member in 2000, but lost all its representatives in the 2006 general election. In the second half of 2010, Tradition, Responsibility, Prosperity 09 (Tradice, Odpovědnost, Prosperita 09, TOP09), a new Czech party, applied for full membership in the EPP. The party, partly composed of former members of KDU-ČSL, participated in the Czech government formed in 2010. It is led by Karel Schwarzenberg and became a full member in February 2011.

In Slovakia, the Christian Democratic Movement (KDH) became a full member of the EUCD in 1990, the same year as it was founded. In the first 10 years it was led by Ján Čarnogurský, who was Prime Minister of Slovakia in 1991-1992. During the term of his successor, Pavol Hrušovský, KDH became first an observer, then an associate and finally in 2004 a full member of the EPP, following Slovakia's entry into the EU. Since September 2009, the party has been led by Ján Figel', former member of the first Barroso Commission. Next to KDH, the Party of the Hungarian Coalition (Strana maďarskej koalície–Magyar Koalíció Pártja, SMK-MKP) became an associate member in 2000 and a full member in 2004. The Slovak Christian and Democratic Union (Slovenská demokratická a kresťanská únia, SDKÚ) became an associate member in 2002. The SDKÚ was led by Mikuláš Dzurinda, formerly of KDH. In 2006, the party merged with the Democratic Party (Demokratická strana) into the Slovak Democratic and Christian Union–Democratic Party (Slovenská demokratická a kresťanská

---

[124] Martens, *Europe: I Struggle, I Overcome*, 195.

únia–Demokratická strana, SDKÚ-DS). This party became a full EPP member in 2004. Dzurinda was Prime Minister of Slovakia for two terms between 1998 and 2006. From 2002 to 2006 he formed a government with two other EPP member parties, the KDH and the SMK-MKP. In July 2010, SDKÚ-DS and KDH formed a new government, together with two other parties, led by Iveta Radičová, Slovakia's first female Prime Minister.

In Hungary, the post-war Democratic People's Party – dissolved by the Communists in 1949 – was refounded as the Christian Democratic People's Party (Kereszténydemokrata Néppárt, KDNP). Together with the Hungarian Democratic Forum (Magyar Demokrata Fórum, MDF), led by József Antall, it formed the Hungarian team inside the EUCD. Between 1990 and 1993, Antall was Prime Minister of Hungary, head of a coalition of the MDF, KDNP and Independent Smallholders' Party (Független Kisgazdapárt, FKgP). The FKgP became an associate member in 2000, but by the time the party could become a full member, it had left the EPP. MDF became an observer in 1998, an associate member in 2001 and a full member in 2004. In 2009 its membership came to an end. The KDNP became a full member in 2004. In 2000 the Alliance of Young Democrats (Fidesz) left the Liberal International and became an associate member of the EPP. Its founder and leader, Viktor Orbán, had transformed the party from a radical student movement into a moderate, centre-right people's party. He was elected EPP Vice-President at the 2002 Estoril Congress. Between 1998 and 2002 Orbán led a coalition government of Fidesz, the MDF and the FKgP. In 2003 'Hungarian Civic Union' (Magyar Polgári Szövetség) was added to the party name. It won the 2010 general election with a very large absolute majority and Orbán became Prime Minister again.

For a long time, the EPP was represented in Bulgaria by the Union of Democratic Forces (Săjuz na Demokratičnite Sili, UDFb). The party was created in the immediate aftermath of the Communist downfall. It applied for membership in 1997 and was granted observer status in 1998. In 2007 the party became a full EPP member. Between 1994 and 2001 the UDFb was led by Ivan Kostov, who served as Prime Minister in the period 1997–2001. Kostov resigned as Chair of the UDFb and eventually left the party to establish, in 2004, a new political force, Democrats for a Strong Bulgaria (Demokrati za Silna Bălgarija, DSB). In 2006 the party became an associate member; in 2007 following Bulgaria's entry into the EU, it became a full EPP member. Also, the Agrarian People's Union (Zemedelski Naroden Sajuz, ZNS) – until 2006 known as the Bulgarian

Agrarian People's Union–People's Union (Balgarski Zemedelski Naroden Sajuz–Naroden Sajuz, BZNS-NS) – and the Democratic Party (Demokratičeska Partija, DP) became full EPP members. In 2006, a new party on the centre-right was formed, the Citizens for European Development of Bulgaria (Grazhdani za Evropeysko Razvitie na Balgariya, GERB). The party has been led by Boyko Borissov, Prime Minister of Bulgaria since 2009. GERB became a full EPP member in 2008.

In Romania, two parties became EPP observers in 1998: the Christian-Democratic National Peasants' Party (Partidul Național Țărănesc Creștin Democrat, PNȚCD) and the Democratic Union of Hungarians in Romania (Romániai Magyar Demokrata Szövetség, RMDSz). Both parties became full members in 2007, following Romania's entry into the EU. In addition to these two small parties, the EPP established contacts with the Democratic Party (Partidul Democrat, PD). In 2005 observer membership was granted to the PD; in 2006 the PD was upgraded from observer to associate member and in 2007 it became a full member.[125] For a long time Petre Roman played a key role in the party. He was succeeded as Chair in 2001 by Traian Băsescu. Băsescu won the presidential elections in 2004; in 2009 he started his second term. At the end of 2007 the PD merged with the Liberal Democratic Party (Partidul Liberal Democrat, PLD) and created an even broader centre-right party, the Democratic Liberal Party (Partidul Democrat–Liberal, PD-L).

In the former Yugoslavia, as well, the EPP was eager to find partners. In Slovenia, the first openly non-Communist political organisation had been founded in 1988: the Slovenian Peasant Union (Slovenska Kmečka Zveza). Later on it was renamed the Slovenian People's Party (Slovenska ljudska stranka, SLS). The Slovene Christian Democrats (Slovenski Krščanski Demokrati, SKD) were led by Alojz Peterle. In 2000 the SKD merged with the SLS. Shortly after its unification with the Christian Democrats, Peterle, who chaired the first free Slovenian government between 1990 and 1992, left the SLS and joined the New Slovenia–Christian People's Party (Nova Slovenija–Krščanska ljudska stranka, NSi-KLS). He became an MEP for the NSi-KLS, as well as EPP Vice-President (from

---

[125] Martens, *Europe: I Struggle, I Overcome*, 203: 'This transformation was, politically, a striking moment: it was the first time that a political party with post-Communist origins had adopted the basic principles and programme of European Christian Democracy.'

2006 until 2009) and member of the Presidium of the European Convention, the only one from Central and Eastern Europe. The Slovenian Democratic Party (Slovenska demokratska stranka, SDS), known until 2003 as the Social Democratic Party of Slovenia (Socialdemokratska stranka Slovenije), was founded in 1989. After several splits, Janez Janša became Chair in 1993. Between 2004 and 2008 he served as Prime Minister of Slovenia, leading a coalition government of EPP member parties: the SDS, NSi-KLS and SLS. These three parties became associate members in 2003 and full members in 2004, following Slovenia's entry into the EU.

In Croatia, Ivo Sanader, Prime Minister from 2003 until 2009, transformed the Croatian Democratic Union (Hrvatska demokratska zajednica, HDZ) from a nationalist party dominated by Franjo Tuđman into a modern, pro-European centre-right party. In 2002, the party became an EPP observer. In 2004, the status of the HDZ was upgraded from observer to an associate member. Since 2009 the party has been led by Jadranka Kosor. Along with the HDZ, the Croatian Peasant Party (Hrvatska seljačka stranka, HSS) and the Democratic Centre (Demokratski centar, DCc) became observers in 2002. In 2007, the HSS moved from being an observer to an associate member. In FYROM (Former Yugoslav Republic of Macedonia), another EU candidate, the Internal Macedonian Revolutionary Organisation–Democratic Party for Macedonian National Unity (Vnatrešna Makedonska Revolucionerna Organizacija–Demokratska Partija za Makedonsko Nacionalno Edinstvo, VMRO-DPMNE) became an observer in 2007. Party founder, leader and Prime Minister (1998–2002) Ljubčo Gorgievski was eager to establish relations with the EPP. His successor is Nikola Gruevski. In 2010 the VMRO-DPMNE became an associate member. As far as Bosnia Herzegovina is concerned, in 2004 observer status was granted to three Bosnian parties: the Party of Democratic Action (Stranka demokratske akcije, SDA), the Party of Democratic Progress (Partija demokratskog progresa, PDPbih) and the Croatian Democratic Union of Bosnia and Herzegovina (Hrvatska demokratska zajednica Bosne i Hercegovine, HDZBiH). In Serbia, relations have been contentious because of the country's critical, often hostile, attitude towards the EU. Nonetheless, the EPP established contacts with the Democratic Party of Serbia (Demokratska stranka Srbije, DSS). The party was founded in 1992 and its leader ever since has been Voiislav Koštunica, the last President of the Federal Republic of Yugoslavia, serving from 2000 to 2003. He also served two terms as the Prime Minister of Serbia, from 2004 to 2007 and from 2007 to 2008. The DSS, together with G17 Plus, also founded in 1992, became EPP observers in 2003. The Alliance of Hungarians in Vojvodina (Savez vojvođanskih Mađara–Vajdasági Magyar Szövetség, VMSZ) became an EPP observer party in 2007.[126]

The EPP strongly supports the inclusion of the Western Balkans in the EU. This does not, however, only involve the successor states of former Yugoslavia. It also includes Albania. The Democratic Party of Albania (Partia Demokratike e Shqipërisë, PDa) was founded in 1990 by Sali Berisha and Genc Pollo. It was the first opposition party under the Communist regime and also ruled the country between 1992 and 1997. In 1999 Pollo left the PDa to establish the New Democratic Party of Albania (Partia Demokrate e Re). This party and the PDa merged again in 2008. Since 1997 the party has been chaired by Berisha, President of Albania between 1992 and 1997 and Prime Minister since 2005. In 2003 the PDa was granted EPP observer status.

In Estonia the EPP established relationships with two parties that then merged in 2006. The Pro Patria Union (Isamaaliit) was founded in 1995 by merging the Estonian National Independence Party (Eesti Rahvusliku Sõltumatuse Partei) and the National Coalition Party 'Pro Patria' (Rahvuslik Koonderakond 'Isamaa'). Its party leader, Mart Laar, was Prime Minister of Estonia from 1992 to 1994 and from 1999 to 2002. In 1998 the party was granted EPP observer status. Another centre-right party, the Res Publica Party (Erakond Res Publica) was founded in 2001. Its leader, Juhan Parts, was Prime Minister between 2003 and 2005. In 2003 the party became an associate member of the EPP. Both parties, the Pro Patria Union and the Res Publica Party, became full EPP members in 2004, following Estonia's entry into the EU. In 2006 they merged into the Union of Pro Patria and Res Publica (Isamaa ja Res Publica Liit, IRL). Since 2007 the new party has been chaired by Mart Laar.

In Latvia the EPP is represented by three centre-right parties. The New Era Party (Jaunais Laiks, JL) was founded in 2001. After having won the 2002 legislative elections, it formed a coalition government led by party leader Einars Repše as the Prime Minister. Except for some months in 2004, he remained Prime Minister until 2006. In 2009 the New Era Party

---

[126] In Montenegro there is currently no EPP member party. A fact-finding mission to Kosovo, organised by the Working Group 'EPP Membership', has been postponed because not all EU Member States have recognised its independence from Serbia.

regained power and Valdis Dombrovskis, who served as Finance Minister from 2002 to 2004, became Prime Minister. The People's Party (Tautas Partija, TP) was founded in 1998. Between 2004 and 2007 the party led the country with Aigars Kalvītis as Prime Minister. Except for the period 2007–2009, since 2002 both parties have stayed in government, as either senior or junior partner. Both parties became full EPP members in 2004, following Latvia's entry into the EU. In 2010 the Civic Union (Pilsoniskā Savienība, PS) became a full EPP member as well.[127]

In Lithuania a Christian Democratic party was re-established in 1989. It was called the Lithuanian Christian Democrats (Lietuvos krikščionys demokratai, LKD) and dated back to the early twentieth century. LKD performed well in the 1992 general elections, and in 1996 it entered into a coalition with the Homeland Union (Tėvynės Sąjunga, TS) and formed a new government. However, the coalition broke up in 1999. The TS was founded in 1993 by Vytautas Landsbergis, the first head of state following the country's independence. As Lithuania became an EU Member State in 2004, both LKD and TS became full EPP members. In 2008 the TS merged with the Lithuanian National Union (Lietuvių Tautininkų Sąjunga, or Tautininkai) and LKD, after which the party name became the Homeland Union–Lithuanian Christian Democrats (Tėvynės Sąjunga–Lietuvos krikščionys demokratai, TS-LKD). In this way, two EPP member parties joined forces successfully. Its current leader, Andrius Kubilius, has served as Prime Minister since 2008.

In Eastern Europe the prospect of EU membership is more remote, and a lot is still to be done to establish stable democratic party systems. However, this has not prevented the EPP from taking concrete steps to find allies for the realisation of its political project. In Belarus, for instance, two parties were granted observer status in 2006: the Belarusian Popular Front (Bielaruski narodny front, BPF) and the United Civil Party (Abjadnanaja hramadzianskaja partyja Bielarusi, UCP). In Ukraine, the People's Union Our Ukraine (Narodnyi Soyuz Nasha Ukrayina, NSNU), the party of Viktor Yushchenko, President between 2005 and 2010, became an observer in 2005, along with the People's Movement of Ukraine (Narodnyi rukh Ukraïny, RUKH). In 2008 also, Yulia Timoshenko's Fatherland Party

---

[127] Sandra Kalniete, Latvia's first European Commissioner in 2004, joined the New Era Party in 2006. In 2008 she left and chaired the newly founded PS. Since her election as MEP she has joined the EPP Group in the EP.

(Batkivshchyna) became an EPP observer. The EPP's efforts to end the discord between the forces that oppose the Communists have, so far, been in vain. In Moldova, the Christian Democratic People's Party (Partidul Popular Creştin Democrat, PPCD) obtained observer status in 2005. The Liberal Democrat Party (Partidul Liberal Democrat din Moldova, PLDM) applied for observer status with the EPP in 2009. This was accepted in February 2011. Since 2009, party leader Vlad Filat is the Moldovan Prime Minister. In Georgia, the United National Movement (Ertiani Natsionaluri Modzraoba, ENM) was accepted as an observer in 2008. Party leader Mikheil (Misha) Saakashvili has been President of Georgia since 2004. Further south, Turkey, an official EU candidate, also has a party with a clear EPP affiliation. The Justice and Development Party (Adalet ve Kalkınma Partisi, AK Parti or AKP) has ruled the country since 2002 with Recep Tayyip Erdoğan as Prime Minister. In 2005 the party was granted observer status.[128]

Prospects

Since 1989 almost 50 parties from Central and Eastern Europe have joined the EPP, be it as observer, associate or ordinary member. Half of them are ordinary members and consequently send their MEPs to the EPP Group in the EP. Enlargement is therefore a big success for the EPP and its Group. The party showed an enormous integration capacity, while at the same time staying faithful to its basic ideas and values. As the EPP is now represented in almost all European countries, membership applications have decreased considerably. The job is not finished, however. Many parties are still waiting to become ordinary members. But their future does not lie entirely in the hands of the EPP. First of all, the countries of these parties have to become candidates for EU membership – for the parties to be granted associate membership – and later on EU Member States – for them to become ordinary members of the EPP. The EPP is in favour of including the Western Balkans in the EU. The party also supports initiatives to establish sustainable relationships with countries such as Turkey, Ukraine and Belarus. Enlargement fatigue in many of the EU Member States, however, has made things a lot more difficult than they were a couple of years

---

[128] Wilfried Martens, *Europe: I Struggle, I Overcome*, 217–20.

ago. At the same time, much still needs to be done in the countries involved. Internal stability is not always guaranteed and some countries are still coming to terms with their past. As with previous waves of enlargement, the EPP is a strong player, dedicated to assisting parties and supporting change in Eastern Europe.

Chapter Five

Merger with the EUCD and EDU

After the European People's Party (EPP) was founded and met with initial success, the European Union of Christian Democrats (EUCD) began a drift into irrelevance which had become obvious by the beginning of the 1980s. Diogo Freitas do Amaral, then EUCD President, took a first step towards merging the two organisations. However, the impulse, which sought to include the globally active Christian Democrat International (CDI), was far too ambitious, and faltered. In 1982 Freitas do Amaral and, soon afterwards, Secretary General Giuseppe Petrilli resigned for different reasons, with no succession having been considered. That signalled a crisis which could not entirely be concealed by the energetic and determined leadership of former EUCD President Kai-Uwe von Hassel, who took over the leadership in his capacity as First Vice-President. EUCD activity had for a considerable period amounted to no more than organising meetings for its constituent bodies.

Merging the Secretariats and Integrating the Executive Bodies

The first step in resolving the situation was to make the same person Secretary General of both the EPP and the EUCD, and as a logical extension, to fuse the secretariats of the two bodies. Thomas Jansen, who had been appointed Secretary General of the EPP in April 1983, was elected Secretary General of the EUCD a few months later. The EUCD's General Secretariat in Rome was dissolved and its functions taken over by the EPP's General Secretariat in Brussels.

Jansen's intention was to unite the leadership of both organisations, politically and in terms of resources. At the same time, efforts were to be made to clarify what each of the two federations should contribute. Both in organisation and function, the two were very different. The EPP was a supranational, federative party active in the political system of the European Community (EC), taking on responsibilities at both parliamentary and executive levels. The EUCD was the organisation of Christian Democracy, in principle uniting all the members of this political family from every country in Europe.

The single leadership was supposed to ensure that Christian Democratic parties in Europe followed a coherent political line, irrespective of whether their field of activity was in the EC, the Council of Europe (CoE) or other European or international bodies. It was also hoped that merging the secretariats would help ensure that the expectations, points of view and interests of various non-EC countries, would be taken into account when decisions were made in the EPP. The reverse held true for EPP decisions: the goal was to formulate policies appropriate to the whole of Europe that did not lose sight of the EC perspective. Joint meetings of executive bodies and the forming of joint committees and working groups served the same purpose.

A special effort was needed to keep alive the EUCD's sense of its own value in relation to the EPP, since the EPP was increasingly important within the EC. This was not just a matter of justifying the existence of the EUCD to members who were also EPP members, and who thereby gained political advantages that made the EUCD less relevant to them. The EUCD was also supposed to be revived as a forum for action and cooperation for those not in the EPP and those who felt the EPP's neglect of them was discriminatory.

Impulse and Motives for Merging the Organisations

The issue of merging the two groups resurfaced when Spain and Portugal joined the EC in 1986. The effect would be to increase the number of parties in the EPP, as well as to further deplete the ranks of EUCD. This time the prospect was of another round of EC enlargement, heading in the direction of including every country in Western Europe. Fourteen of the EUCD's 21 member parties – in other words two-thirds of its members, among them the largest and most influential – were already in the EPP. Under the circumstances, it no longer seemed sensible to retain the two organisations, especially given the joint General Secretariat.

Experience had shown that integrating the EPP and EUCD (while retaining the fiction that they were distinct) created no political problems. In particular, no attempts were made by either a single member party or a group of EUCD parties to undermine or dilute any position taken by the EPP, whether in relation to further development of the EC or to Community policy. Even the seven EUCD parties that could not join the EPP did not insist on keeping the EUCD as a distinct organisation. They were interested in having as close a relationship as possible with the parties of the EC and in adopting for themselves rules similar to those that already existed in the EPP.

Another factor was at work: the coexistence of two organisations at European level muddled the public image of the movement and sapped confidence in its ability to act. The image problem, made worse by the parallel existence of the European Democratic Union (EDU), was not resolved by the 1983 merger of the EUCD and EPP secretariats. Neither was the inefficiency resulting from duplication. These headaches would remain as long as there continued to be two (or even three) presidents, presidencies and executive bodies required to arrange separate events for each organisation, not to mention the need for separate representation and financing.

Initiative, Debate and Decision-Making

In 1984 the European Union of Young Christian Democrats (EUYCD), a recognised association of the EPP and of the EUCD, had made a powerful case for merging their union with two previously coexisting federations. They pushed for their parent organisations to follow suit. They made the point that the united Europe for which the Christian Democrats were striving had also had to overcome differences before achieving integration – here the EC, there the CoE and so on. At the EUCD Madrid Congress in 1985, the EUYCD formally proposed that 'attempts should continue to be made to combine, more than at present, the political and technical work of the EUCD and the EPP'.[129] This was unanimously accepted and led to discussions within the EUCD and EPP. Initially these bore no fruit. But a year later, at the EPP Congress in The Hague, the EUYCD once again put forward a proposal. This time it explicitly demanded the fusion of the EUCD and EPP:

> The VI Congress of the EPP requires the Secretary-General of the EPP and EUCD, in agreement with the Presidents of both organisations, to present to the EPP Political Bureau, before the end of 1986, a proposal for changes to the EPP statutes which would above all allow for admitting to membership parties from outside the EC, and prepare the way for the complete fusion of both organisations, while taking into account the wishes of all member parties, and of EPP work specifically developed for the Community.[130]

The motion was rejected, but it was passed on to the Political Bureau of the EPP and of the EUCD for further consideration. Both Presidents and the Secretary General were assigned the task of taking soundings and then proposing how to proceed. These soundings, and detailed discussions over the course of 1986 and 1987, revealed a general agreement with the project, though a number of party representatives tended to be sceptical about the terms.

Passage of the resolution was held up by opposition from the EUCD President and the Italian delegation for reasons related to political issues and to personalities. The Italians were concerned about the identity of the EPP as a party federation relating to the EC. They also assumed that the fusion of the EUCD and EPP would lead to the demand, especially by the Germans, to admit the Conservative parties, something they had always refused to do. They no doubt also feared losing the post of president, occupied since 1985 by Emilio Colombo (and before that, from 1983, by Guilio Andreotti).

However, in the summer of 1988 a committee was finally formed to look into the merits of merging the EUCD and EPP and to assess other methods of strengthening Christian Democracy as an active force in Europe. The committee was also asked to draft a statute for a renewed federation of European parties that would be capable of taking effective

---

[129] Decision by the Twenty-second EUCD Congress in Madrid, 1985, EUCD General Secretariat Archive (Brussels).

[130] Resolution of the EUYCD for the Sixth EPP Congress in The Hague, 10–12 April 1986, EPP General Secretariat Archive (Brussels).

action.[131] The Secretary General was to chair the committee, which was to comprise a representative from each member party, a representative of the Christian Democratic Group in the Parliamentary Assembly of the CoE, a representative of the EPP Group in the European Parliament (EP) and representatives from both the EUCD and EPP Presidencies. The results were to be presented for decision by spring 1989 at the latest.

A 'Large' EPP

The committee's mandate was kept very general, out of sensitivity to the Italian delegation's reservations, but it did amount to creating a new structure for a 'large' EPP, in which all European parties with a Christian Democratic orientation could cooperate as equals. There had for some time been an understanding that integration would take place on the basis of the EPP statutes, meaning the EUCD would be absorbed by the EPP. It was understood that this would be effected by absorbing the last non-EPP members of the EUCD. To do this, the EPP statutes would have to be changed in two respects: conditions for membership and voting rights. The opportunity would also be taken to embark on a radical revision of the 1976 statutes.

In accordance with earlier EPP decisions, this reform was to be made in light of 'the perspective opened up by the Single European Act', as well as 'the experiences and achievements of cooperation within the EUCD and the EPP', and finally in view of 'the political developments of the European Community since direct election of the European Parliament'.[132] Such ideas gave rise to proposals for a new set of statutes. The proposals were repeatedly debated by the Political Bureau of the EPP and of the EUCD and in principle approved by them. Given that parties from countries which did not belong to the EC had the right to vote for the EPP leadership, for instance, the proposal envisaged that such parties should take part in all discussions and on all matters. However, they would not be allowed to take part in decision-making about the constitutional development of the Community or on Community political issues generally.

---

[131] This was the mandate of the Statutes Committee, decided at the EUCD and EPP Political Bureau meetings on 7 June 1988, EPP General Secretariat Archive.

[132] Ibid.

The period leading up to the final discussion of the proposed statutes co-incided with the revolutionary events taking place in Moscow and in other capitals of Eastern Europe in the late 1980s. Suddenly the possibilities for cooperation across the whole of Europe opened up. Given the new situation, the EUCD Political Bureau decided in spring 1989 to postpone the merger project. Instead, a radical reform of the statutes, strengthening both the EUCD's independence and its capacity to act, was designed to prepare that organisation to welcome the Christian Democratic forces of Central and Eastern Europe as soon as they had succeeded in forming themselves into parties. These forces were now free again and were taking the first steps to organise themselves, setting up groups, clubs and forums in preparation for founding parties in several countries.

The looser and less binding EUCD framework seemed more suited than that of the EPP to deal with the expected influx of new parties from countries which so far lacked political systems based on the rule of law and democratic structures. In any event, from the EPP's point of view it was quite unthinkable to take in as members or associate members parties from Eastern and Central Europe about which nothing was yet known. Furthermore, the countries in which these new parties were preparing to take over political responsibility would have no chance, in the short or medium term, of joining the EC. All this brought to an end the mandate to effect a merger of the EUCD and EPP. The committee now set about using the work it had done to develop new statutes for both the EPP and the EUCD.

The New EPP Constitution

The revision of the EPP's statutes had, for some time, a momentum of its own, independent of the merger project. On 2 December 1983, immediately after taking office, Secretary General Thomas Jansen wrote a memo identifying the source of a number of deficits in EPP working methods and the ways in which its various bodies saw themselves. These were the result, he said, of structural problems which were especially revealed in the fact that there was a striking gap between the demands of the EPP statutes with respect to the composition and role of the Political Bureau and reality. This circumstance in turn corresponded to the gap between the pretension evoked by the name 'European People's Party' and the association's actual circumstances: it remained, as ever, not much more than a 'Federation of

European Community Christian Democratic Parties'. Secretary General Jansen wrote: 'I do not want to draw the conclusion from this that we should withdraw this pretension. On the contrary. We should fortify it by taking in the fact that we can only reach the goals embodied in the name "EPP" if we don't behave as if we had already reached them.'[133]

This memo in turn gave rise to proposals for a careful medium-term reform of the statutes, to '[b]ring the working methods of the EPP bodies into line with the actual state of the EPP's development as a federation of national parties' and to be able 'by these means to avoid many a frustration produced by exaggerated expectations, and many an absurdity created by the distance between pretension and reality, and eventually to get better political results'.[134]

The Secretary General was given the task of placing before the party's responsible bodies, by the end of 1984, proposals for constitutional reform,[135] but the task was quickly overtaken by the discussion about merging with the EUCD, which soon afterwards blotted out everything else. Reflections on how to revise the statutes were swamped by the debates about a new constitution for the 'large' EPP. Those, in turn, were connected with a proposed revision of the EPP statutes being developed by a working group chaired by the Secretary General from spring 1989 onwards. The text was discussed and voted through at the Dublin Congress in November 1990.

The 1990 revision of statutes introduced a series of important innovations. The subtitle of the party's name, correct until now, but clumsily expressed as 'Federation of Christian Democratic Parties of the European Community', was simplified, and the name of the party was now officially 'European People's Party: Christian Democrats'. The member parties were now obliged to 'represent the policies taken up by the EPP in the context of the European Community in their national policies'.[136] The designation 'associate member' applied to Christian Democratic parties from countries in the process of negotiating their entry to the EC. The possibility of attaining the status of 'permanent observer' would be offered to parties of simi-

---

[133] Struktur und Arbeitweise der EVP-Gremien, Memo to the Political Bureau, EPP General Secretariat Archive.
[134] Ibid.
[135] Decision of the EPP Political Bureau, Strasbourg, 26 June 1984, EPP General Secretariat Archive.
[136] Article 3 of the 1990 EPP Statutes.

lar political orientation. 'Individual membership' was also introduced, while the Conference of Party and Government Leaders was recognised as a party body.[137] Next, an elected Presidium replaced the Executive Committee, which had been made up of a large number of ex officio representatives and Presidium members. The Secretary General's role was strengthened and officially recognised. Representation in the various party bodies was to be reorganised to accord with political realities. Conditions were imposed on speaking and voting rights, committees and working groups, associations, the treasurer and finances.

The constitutional reform meant adapting the standing orders, as well as, for the first time, revising the financial regulations (equally applicable to the EPP and EUCD). These revisions, among other things, stipulated the levels of contribution, the conditions for grants to associations, as well as the method of calculating them, and the accounting rules and financial control of the General Secretariat.

The new statutes and the documents relating to how the organisation should be run and financed gave the EPP a constitution that clearly stressed its claim to be a European party. The conditions it contained did not, as some paragraphs of the 1976 constitution had, exaggerate the extent to which the member parties or their representatives could or should be present or participate. Beyond that, the party bodies and system of rules remained flexible enough to adapt to the effects of more intensive political and institutional integration of the European Union (EU) in the years ahead.

A New Role for the EUCD

In spring 1989, the idea of merging the EPP and EUCD had been abandoned. That decision was taken against the background of the revolutionary events in Moscow and the capitals of Central and Eastern Europe. The hope was that the EUCD would play an important, autonomous role in rebuilding democracy and in bringing together Christian Democratic forces and their allies in the countries liberated from Communism.

At the Twenty-third EUCD Congress in Malta in November 1989, Christian Democrats from Poland, Hungary and Estonia were able to take

---

[137] See chapter eight of the present volume.

part for the first time. And they did not represent some exile organisation or another, but parties or groups active in their own countries. The EUCD's Political Bureau, Presidium and General Secretariat were now focused on one goal: the transition from a Western European organisation to one that embraced the whole continent.

The Christian Democratic Union of Central Europe

A conference of representatives from Central and Eastern European parties with a Christian Democratic direction took place in Budapest at the beginning of March 1990. It was convened by the Christian Democratic Union of Central Europe (CDUCE) and organised in conjunction with the CDI and the EUCD and EPP Group in the European Parliament. The parties that were represented used the opportunity to express the hope that they would soon be accepted as members of the EUCD. But they also decided to revive the CDUCE to promote their parties, particularly in the transition period.

The CDUCE had come into being during the 1950s.[138] It incorporated the Christian Democrats of Central Europe, who had been in exile in America or Western Europe and represented the legitimacy and historical continuity of the parties in their home countries, from which they had been forced to leave as a result of Communist repression. During the 1950s and 1960s, the CDUCE, which was also recognised as a regional organisation of the CDI, became very active. It cultivated contacts with democratic opposition groups working underground (or condemned to inactivity) in the Eastern bloc countries. In the West it tried to ensure that the destiny of the people living under Soviet or Communist rule was not forgotten.

Several factors led to the progressive decline in the importance of the CDUCE after the 1970s, even within the international Christian Democratic movement. The politics of détente made the relevant political forces in the West more indifferent, and the advancing age of the leading figures was also a factor.[139] Despite the best intentions of its protagonists, the

---

[138] See chapter one of the present volume and also Roberto Papini, *L'Internationale démocrate chrétienne: la coopération entre les partis démocrates-chrétien de 1925 à 1986* (Paris, 1988).

[139] The leading representatives of the CDUCE were Konrad Siniewicz (Poland), Secretary General, and Bohumir Bunza (Czechoslovakia), President, who by

CDUCE entered the new setting of a liberated Central Europe without the necessary organisation, people and ideas to play a decisive role. Further, those who had been responsible for the re-establishment and development of parties after the fall of Communism in 1989 were not interested in having their own Central European party federation. They wanted to belong to an organisation that included all of Europe, namely the EUCD, which could guarantee direct cooperation with the Western European parties and the EPP in particular. They did not want to take orders from the 'old men' (however highly respected) who had returned from exile.

Therefore, in the course of the Twenty-fourth EUCD Congress in Warsaw (21–23 June 1992), the representatives of the new Central European parties resolved to disband the CDUCE. That step also freed them from the pretensions of their compatriots in exile to speak for them.

What the EUCD Tried to Do

From the outset, the EUCD had worked hard to establish contacts and cooperation with the Central and Eastern European Christian Democratic parties and support them as they gradually came back to life. Increasingly, the parties' representatives were included in the work of the EUCD. The starting point was a conviction about Western efforts in general, and EC policy in particular, towards the countries of Central and Eastern Europe: that in the short term, efforts must be concentrated on establishing democracy and a social market economy. In the medium term, the goal must be economic improvement, social justice, and social and cultural recovery; the long-term aim, membership in the EC. From autumn 1989 onwards, a special EUCD and EPP working group more or less systematically compiled information and coordinated projects and activities. The group was chaired by Karl Josef Hahn and, from 1990, by Wim van Velzen.

Over the years, many EUCD member parties arranged opportunities for further meetings and exchanges. It is also worth noting the keen efforts of the foundations and institutes connected with the Christian Democratic movement. Like-minded people from both parts of Europe were given the oppor-

---

1990 were over 70 years old. They were replaced by Sandor Karczay (Hungary) as President and Ivan Carnogursky (Slovakia) as Secretary General. All the other members of the new executive were representatives of governments in exile and lived in Rome, Paris, Brussels, Geneva, etc.

tunity to meet and to debate in a range of seminars. The Political Academy of the ÖVP, the German Konrad Adenauer and Hanns Seidel Foundations, the Dutch Edoardo Frei Foundation in conjunction with the Scientific Institute of the Dutch Christian Democratic Appeal (Christen Democratisch Appèl, CDA), the Belgian Centre for Economic, Political and Social Studies (Centrum voor Economische, Politieke en Sociale Studies, CEPESS), the Italian Fondazione Alcide de Gasperi and the Paris-based Fondation Robert Schuman all did extremely important work in this connection.

The EUCD itself organised a number of seminars and colloquia for representatives of Central and Eastern European parties. It also established the Academy for Central and Eastern Europe in Budapest, offering systematic education and training seminars to current and future party workers in Christian Democratic parties and those politically allied to them. Finally, there were many visits to Central and Eastern Europe by all sorts of delegations. These took place under the auspices of the EUCD, the CDI, the EPP Group, or individual member parties and their parliamentary groups. The delegations wanted to learn about political developments and the problems faced by politically friendly parties. This intensive travel, which included attending party Congresses held by the new parties and participating in their election campaigns, had two essential aims. One was for Western parties to make themselves a familiar part of the landscape, and the other, to demonstrate the solidarity of the Western European Christian Democrats.

The EUCD's New Constitution

The EUCD's structures needed to be adapted to the new political situation in Europe. The constitution had to be reformed, and that reform needed to include new rules for granting membership. Moreover, voting rights and other matters related to membership had to be reorganised in light of what would likely be a huge increase in the number of members. The EUCD's organisation as a whole needed to be modernised and adapted to current and future challenges, taking into account the EUCD's experiences since it was founded in 1965. The EUCD had to redefine its own value in relation to the EPP, and to decide anew on how to divide the work between them.

After long discussions in a committee chaired by Secretary General Jansen, the Political Bureau finalised the new EUCD constitution in autumn 1991. It came into effect at the same time as the Warsaw Congress in June the following year.[140] On the basis of this new set of rules, the EUCD Council (representing member parties) met in January 1993, elected a new president and laid down the general direction for future work. Colombo, President since 1985, was succeeded by Wilfried Martens, who was now President of both the EUCD and the EPP. For the first time, representatives of Central European parties were elected to the Presidency: Alojze Peterle (Slovenia) and László Surján (Hungary).

The Warsaw Congress clearly marked a break in the EUCD's history, which at this time united 28 member parties and 11 further parties with observer status. It was the first time an EUCD Congress had taken place in the heart of Central Europe and the first time that representatives of Christian Democratic parties from the whole of Europe had been represented. And it was the last time a Congress was to take place on the basis of the constitution agreed to in 1965 when the EUCD was founded. Up until the peaceful revolution in Central and Eastern Europe, the EUCD had remained a purely Western European affair. Now, at last, it had become what its name claimed it to be: the European Union of Christian Democrats.

New Parties Join

In deciding on membership applications, the relevant EUCD bodies always sought detailed information and made sure there was consensus among member parties about whether or not to admit a new member. It was clearly essential to have a basis for deciding on the admissibility of a party, and this had to be as unambiguous as possible. Not only did many parties want to join the EUCD, but there was considerable political pressure from different directions. There were constant real and pressing reasons from this or that (Western European) party as to why certain (Central or Eastern European) parties should immediately be admitted or given precedence over others. Often, several motives were in play at the same time: party/political, governmental, national/sentimental, ideological, personal and tactical. But given competing interests, such considerations were usually not likely to find general agreement, which was why the Central and Eastern

---

[140] EUCD statutes of 1992, EUCD General Secretariat Archive.

Europe Working Group defined criteria for admitting members, criteria later enshrined in the EUCD constitution.

Recognition of observer status was granted to parties which had not (yet) fulfilled the requirements for acceptance as members, or where doubts existed about their identity, development, role or politics. Observer status was above all justified by the extent to which it served the political or operational interests of the EUCD, whose goals were to make sure its message was heard in as many countries as possible and to develop appropriate political structures.

This task also involved a certain pedagogical responsibility on the part of the EUCD towards parties wanting to join. Such parties had to be given the chance to learn by doing, by taking part in events. Active observation would also teach them how to develop their political programmes, their instruments of power and their whole modus operandi, all with the eventual object of becoming members. Observer status implied a right to political support. Opening up these possibilities was eventually a natural part of showing solidarity, even an obligation. In return, the parties committed themselves to Christian Democracy and put their hopes in the EUCD. There were no institutional links between the EUCD and the observer members, and the only obligation the EUCD took was to invite their representatives (without the right either to speak or to vote) to EUCD events. All decisions which subsequently had to be made about admitting new members or recognising them as observers were based on these criteria.

In each case, thorough investigations took place in Central and Eastern Europe, carried out by specially commissioned delegations, and there were detailed discussions in the Working Group and in the EUCD Council. Even so, not every decision reached turned out to be tenable. The party landscape during those first years of democratisation was shifting so quickly that a number of changes had to be made. The Hungarian Small Landowners Party (Független Kisgazdapárt, FKgP) lost its membership, and the Polish Labour Union (Stronnictwo Pracy) had its observer status withdrawn. The status of several other parties needed re-examination because of developments either in domestic politics or in the parties themselves.

There was, moreover, no active political recruitment drive by the EUCD among potential member parties. The formal criteria for a membership application are one thing, but the most important sign of a party belonging to the Christian Democratic family must be its political identity and what logically follows from that: the party's readiness to take the initiative. In light of EU expansion to bring in the countries of Central and Eastern Europe, all EUCD members became candidates for EPP membership. The EPP constitution allowed for associate membership for those parties whose countries' EU membership was being negotiated, and several EUCD member parties from Central and Eastern Europe made such applications.

Integration of the EUCD into the EPP

All this raised the issue, now much more urgent, of whether member parties could continue to afford the luxury of two overlapping structures, the EPP and the EUCD. It soon became clear that most EUCD member parties, for very sound reasons, wanted to join the EPP. The Western European member parties had progressively lost interest in the EUCD, and the political rationale for these parallel structures became ever more dubious. Now that the countries of Central and Eastern Europe had a realistic chance of joining the EU, the arguments of spring 1989, which had persuaded the EUCD and EPP to postpone indefinitely a fusion of the two organisations, were no longer convincing.

But there was still resistance to a merger. Klaus Welle, Secretary General of both the EPP and EUCD, whose determined efforts to unite the two organisations in one body were supported by the main party leaders, described the situation:

On the one hand there were justifiable human interests and problems. The EUCD offered many a field of action, whether as presidency member or as the representative of a member party or association. All this falls away once you have integration in another solid framework. The EUCD can also, after all, look back on a history of 50 years, the first 20 under the name of the Nouvelles Équipes Internationales (NEI). Some who are still active today were there when it was founded. Those Western European parties which were still rather reticent about the EPP's integration policy (Kristelike Folkeparti Norway, the Swiss Christlich-demokratische Volkspartei) had similar reservations to those Central and Eastern European parties whose countries had not yet tried to join the European Union. Moreover, the Central and Eastern Europeans who were full EUCD members posed the question of what status they would be accorded in the EPP. Moreover, the parties which did not belong to the EDU wondered whether the fusion of the EUCD with the EPP did not, in effect, force Central and Eastern European parties to join the EDU, something which in turn would have meant a growing distance from Christian Democracy. Finally, one should not neglect to mention those who saw EU enlargement as a danger to deepening cooperation in the Union, and who for that reason disputed the sense of integrating Central and Eastern European parties into the EPP ahead of such enlargement.[141]

Despite the scepticism and the doubts that surfaced again at the EUCD Council in Oslo in February 1996, it was nevertheless decided to raise the issue at the next EUCD Congress. EUCD Vice-President van Velzen, who was preparing to take over as President from Martens, was given the task, along with Secretary General Welle, of elaborating a working plan for the years to come, including a timetable for integrating the EUCD into the EPP. The document produced for the EUCD Council meeting on 14–15 June proposed integration by 2002 at the latest. EPP committees were urged to speed up the process by offering the parties from Central and Eastern Europe concrete possibilities of working with them.[142] A little later, on 10–11 July, the EPP 'mini-Summit' and Political Bureau endorsed the strategy set out in the document and decided to start the appropriate processes and changes to the statutes.

The Twenty-fifth EUCD Congress took place in the Slovene capital, Ljubljana, on 4–6 October 1996. Following intensive debate and some amendments, delegates unanimously voted for the proposed integration that had been agreed upon by the Council on 12 September. Van Velzen chaired the EUCD's last Congress.

Apart from issues relating to future cooperation between Christian Democrats in a Europe now in the process of reuniting, the EUCD Congress also dealt with a comprehensive document entitled 'Stability, Security and Prosperity in the New Europe'. This was a synthesis of the Christian Democratic programme on fundamental questions of political ethics and democracy, as well as an overview of various practical European political issues: internal and external security, minority rights, the social market economy (its foundations and current challenges), employment, family

---

[141] Klaus Welle, 'Reform of the European People's Party 1995–1999', in Hans-Joachim Veen, (ed.), *Christlich-Demokratische und Konservative Parteien in Westeuropa 5,* Studien zu Politik, vol. 31, Konrad Adenauer Foundation (Paderborn, 2000), 543ff.

[142] Documents relating to the integration of the EUCD into the EPP can be found in the *EPP/EUCD Yearbook 1996*, published by the EPP Group in the EP (Brussels, 1997), 318–33.

policy, environment and energy. This programme was designed to serve as the basis for new and future forms of cooperation.

The EPP Congress in Toulouse on 9–11 November 1997 was now called on to adopt the necessary statute changes which would finally accomplish the integration of the EUCD into the EPP. To make it easier to unite the two organisations and enable them to work together, the hurdles in the new statutes were substantially lowered. A party that fulfilled other conditions and had established itself as a national political force would now be granted associate membership if the country involved had applied for EU membership.

This made it almost automatic that EUCD member parties would become EPP members once their countries had applied to join the Union. The EUCD thus became redundant, and its dissolution during the next EPP Congress, scheduled for spring 1999, ahead of the June European elections, was agreed upon. There were two final meetings of the EUCD Council in 1998, in Sofia on 7–8 April and in Madrid on 13 November. The main subject on the agenda at both was once again the process of integration into the EPP. In Madrid the process of integrating the EUCD, and therefore its dissolution, was sealed, provided the necessary formal decisions were reached at the next EPP Congress on 4 February 1999.

After all the wearisome consultations, the complex process of building a consensus over many years and, at last, the unanimous agreement, it was hardly to be expected that the EPP Congress would take a different direction. There was agreement on overcoming the last hurdles; EUCD President van Velzen delivered a final report to the Congress. The European Union of Christian Democrats, from which the European People's Party had originally sprung, was no more. Its inheritance passed to the EPP.

The EDU and EPP

In parallel with the 'identity crisis' of the EUCD, the EDU also lost relevance as member parties from East and West joined the EPP because of developments in the EU.[143] EPP membership would give them greater in-

---

[143] See Andreas Khol and Alexis Wintoniak, 'Die Europäische Demokratische Union (EDU)', in Hans-Joachim Veen, *Christlich-Demokratische und Konservative Parteien in Westeuropa 5*, 403–57.

fluence. What, now, was to become of the EDU? Alois Mock from Austria had been the EDU's charismatic Chair and backer for 19 years. After he retired in 1997, he was replaced by Andreas Khol, also from Austria, who had worked closely with him as Executive Secretary since 1994, infusing the organisation with practical and intellectual dynamism. But now the EDU lacked the dynamic forces that might have secured an independent role for it in the new Europe.

Sauli Niinistö (from Finland) as Chair and Alexis Wintoniak (from Austria) as Executive Secretary, for all their considerable commitment and effort, did not play the same role as had Mock and Khol, who had founded and essentially shaped the EDU for most of its existence particularly because the whole political scene had shifted. Those strongly interested in retaining the EDU at the end of the 1990s were actually only those EDU member parties that, for political or ideological reasons, did not or could not contemplate eventual membership in the EPP. The best known of such parties were the British Conservative Party, the Czech Civic Democratic Party (Občanská demokratická strana, ODS) and the Turkish Motherland Party (Anavatan Partisi).

For most of the other EDU member parties, including those which had been the most enthusiastic about party cooperation within the EDU framework, the German Christian Democratic Party (Christlich Demokratische Union, CDU), the Bavarian Christian Social Union (Christlich-Soziale Union, CSU) and the Spanish People's Party (Partido Popular, PP), the EDU's usefulness had largely evaporated. What was left was only its function holding together the group of centre-right parties and, specifically, providing a link to the British Conservatives. Those in charge of these parties complained more and more loudly about the waste of time and energy spent holding a multiplicity of meetings and the high financial costs of maintaining two separate structures. The way forward had already been shown in the elegant fusion of the EUCD into the EPP, which in truth amounted to the self-dissolution of the EUCD on condition that its members be guaranteed entry to the EPP. As EPP pragmatists saw it, this was the obvious way to solve the EDU problem.

However, there were substantial differences in the relationships between the EPP and the two other organisations. Most important, there was no institutional relationship between the EDU and the EPP. Over all the years of their parallel existences, such a relationship had been carefully avoided by the EPP. All proposals put to the EPP's committees from the late 1980s about achieving a better division of labour and the synergies that would result had been rejected.

Although a good number of EDU parties were at the same time members of the EPP, the two bodies tended to regard each other as competitors standing for different principles of organisation. Whereas the EDU followed the principles of cooperation and diplomacy, the EPP stood for democracy and federalism. It is fair to say that those EPP parties which had always rejected participation in the EDU also tended to make an ideology of the democratic federalist principle. By contrast, the EDU parties which would have found membership in the EPP difficult or even intolerable tended to make an ideology of diplomacy and cooperation. Depending on their political culture and tradition, the parties located between these two positions never had a problem with dual membership, or at least, not a serious one. For them it was a matter of organisational principles to be applied on the basis of opportunity or practicability.

The two forms of organisation served different purposes. The EPP has sought to be one entity as a party. Its intention was to play a role, based on its political programme, in the politico-institutional EU system and, where possible, at other levels of its federal structure, meaning at the transnational level and also, through its member parties, at national and sub-national levels. By contrast, the EDU's function, even if perhaps not its policy, was, beyond any affiliation with the EU, to ensure lasting cooperation between centre-right parties in Europe where it was impossible to do this within a single party in the expanding EU or in some other form of political entity.

Bringing the EDU and EPP Closer Together

Over the course of the 1990s, the Maastricht, Amsterdam and Nice Treaties marked important stages in the progressive consolidation of a supra- or transnational EU politico-institutional system. This was necessarily accompanied by both democratisation and federalisation, a fact that showed the superiority of the EPP principle. Most EDU parties that had long kept their distance eventually acknowledged this by voting with their feet and joining the EPP. They did so even though at the outset they could not, or did not want to, accept all elements of the EPP's programme, which came from the Christian Democratic tradition and the European Movement.

Since the fall of Communism in Central and Eastern Europe and the dissolution of the East–West conflict, there had been a visible shift from ideological justifications for political positions and programmes to pragmatic arguments. This was seen as a form of 'rapprochement' between the EDU and the EPP, and something on which the EDU Chair Niinistö and his Executive Secretary Wintoniak worked very assiduously and systematically. The new circumstances, the result of globalisation and, given the mess left behind by Communism, the wide range of practical economic, social and organisational tasks to be performed in Central and Eastern Europe, demanded unprejudiced answers.[144]

In 1998, the twentieth year of its existence, the EDU under its new Chair, Niinistö, took the initiative of sorting out its relationship with the EPP. The eighteenth meeting of party leaders was held in Salzburg to mark the 20-year anniversary in April 1998 of the Klessheim Declaration, the founding document of the EDU. Based on precisely this declaration, EDU leaders now called for all Christian Democrats, Conservatives and likeminded parties – literally every EDU and EPP member – to be brought together in a single organisation.

The declaration adopted by party leaders, entitled 'Towards the Majority', had been carefully prepared by the EDU Steering Committee. It took into account all the leaders' different, and contradictory, ideas about how to proceed and how to structure a united organisation. Ideas about which doors this decision might now open ranged from starting afresh and allowing the two organisations to merge, to gradually merging. The resolution included a call for closer cooperation between committees and secretariats, and for continuing dialogue between the organisations about political and structural questions. The EDU chair was to consult national parties and the EPP leadership and report back to the EDU leaders' conference in September.

In June 1998, Niinistö met EPP President Martens, and they agreed that the Presidencies of the two organisations should meet in Cadenabbia, Italy, on 18–19 September. Before that, in Rome on 3–4 September, the EDU Steering Committee set out in writing what seemed possible on the basis of consultations that had taken place in the meantime. The document, entitled

---

[144] Ukko Elias Metsola, *Towards the Majority: An Analysis of the Rapprochement between the European People's Party and the European Democrat Union* (Helsinki, 2000).

'Restructuring the Cooperation of the Christian Democrat, Conservative and Like-minded Parties in Europe', described a road map for what was to follow: gradually combining committees, working groups, delegations and seminars, then the secretariats, and finally the decision-making and ruling bodies.

Although there were positive reactions to the suggestions, including by every single member of the EPP Presidency, nothing could be conclusively decided because the EPP Political Bureau and Council had not yet reached a decision. During the discussions which followed the meeting between Martens and Niinistö, it became clear that there was no chance of securing majority support in the EPP, because an EDU move to adopt the Christian Democratic and federalist political programme (contained in the EPP's 1992 Athens Basic Programme) was not a precondition for the merger.

The situation altered after the European elections on 13 June 1999. The power relationship in the EPP parliamentary Group was changed by the substantial increase in the size of the British Conservative delegation (and of the German CDU). As a result, the Group agreed to change its name to 'Group of the European People's Party (Christian Democrats) and European Democrats'. At the same time, the EDU parties which had not yet joined the EPP, meaning above all the British Conservatives, were accorded a degree of autonomy and thus a different political profile. This also made it possible for the French Rally for the Republic (Rassemblement pour la République, RPR) to allow its deputies to join the newly defined Group.[145]

Even if this did not deliver any kind of formal merger to resolve the problem of the relationship between the EPP and the EDU, for the EDU it amounted – psychologically, at least – to an important step in that direction. But it was also a step which had the effect of mobilising those forces in the EPP that did not agree with the direction. There was widespread unease and resistance, both in the parliamentary group and in a number of member parties.

Following the CDU's defeat in the 1998 Bundestag elections and the resignation of Helmut Kohl, Wolfgang Schäuble had become his successor as CDU leader and had devoted himself intensively to the party's European work. Schäuble now proposed a fresh formula for merging the EDU

---

[145] See chapter four of this volume.

and EPP, one which met the EPP's interests. It provided for the EPP itself to become a member of the EDU. Under this plan, the EPP member parties would withdraw from the EDU, and the EPP would in future represent them in the EDU. The EDU Secretariat in Vienna would then be dissolved and combined with the EPP General Secretariat in Brussels.

The proposal, subsequently modified so that EPP parties were free to choose between being represented by the EPP in the EDU or remaining individual members of the EDU, was intensively discussed by leaders of the two organisations and of their member parties in the second half of 1999. In the end, it was rejected at the EPP Summit in Helsinki on 9 December. Some party leaders who had originally gone along with Schäuble's plan had evidently changed their minds – among them, it seemed, Schäuble himself. They may have had the impression that some EPP parties did not want to accept even indirect membership in the EDU, and that others were unwilling to abandon their traditional EDU membership in favour of the EPP.

The possibility of an agreement, either a mutual fusion or formal merger, was now exhausted. All that was left was technical cooperation in the form of agreed work plans, joint projects and meetings, agreed agendas for otherwise separate meetings of decision-making and leadership bodies, and in particular sharing a joint Secretariat with a common budget. That, eventually, was what was decided by a large majority, after a lively debate in the EPP Political Bureau on 10 February 2000. The decision came in the middle of the turbulent events surrounding the government coalition in Austria of the ÖVP, led by Chancellor Wolfgang Schüssel, and the Austrian Freedom Party (Freiheitliche Partei Österreichs, FPÖ).

Within a year the EDU Secretariat in Vienna had been closed down, and Executive Secretary Wintoniak was sitting at a desk in the EPP General Secretariat offices in Brussels. As per the agreement, he became, following a proposal by EPP Secretary General Alejandro Agag, a Deputy Secretary General of the EPP (while retaining his EDU function), along with Guy Korthoudt (responsible for organisational affairs) and Christian Kremer (responsible for political questions). Wintoniak now came under Agag's authority, including in his EDU capacity.

Closing down its Secretariat seemed to have sealed the fate of the EDU. Even if there was good cooperation, especially at the level of working groups, there was much less need for meetings within the EDU framework, either of party leaders or of the Steering Committee. The 2002 resignation of Wintoniak, who transferred to a new job heading the office of President Khol of the Austrian Parliament, effectively marked the end of the EDU as a distinct entity. This found symbolic expression when the last EDU President, Niinistö, was made Honorary President of the EPP at the EPP Congress in Estoril in October 2002. It was the de facto, if not formally agreed upon, integration of the EDU into the EPP.

In the period from 1994 to 2004, the EPP's development mirrored that of the European Union itself as it expanded and consolidated. The shared success of the EDU, EUCD and EPP in working together using different methods to unite the Christian Democratic, Conservative and like-minded parties had now made the EDU redundant, just as it had the EUCD.

Global Perspective

What is still missing, however, is a strong, united international organisation of the centre-right. Since 1983 the CDI has operated in parallel with the International Democratic Union (IDU), an organisation that assembles right-of-centre parties from around the world, including almost all EPP member parties. Since 2008 the EPP has been recognised as a regional union of the IDU. The mere existence of the IDU next to the CDI is unfortunate, given that the impact of a single organisation would increase considerably.[146] Apart from the multilateral involvement in international politics, the EPP embarked in 2005 on an effort to establish its bilateral international contacts with parties and countries beyond the EU and, more importantly, beyond the European continent. A clear example is the bilateral relationship with the US Republican Party, particularly the International Republican Institute (IRI). Since 2008, a number of high-level political delegations from the EPP and the IRI have met in Washington D.C. and Brussels, and there has been regular cooperation on many issues of common interest.[147] Clearly, this new avenue is a true challenge for the EPP and its future development.

---

[146] Wilfried Martens, *Europe: I Struggle, I Overcome* (Brussels, 2009), 220–1.
[147] Ibid., 221–2.

Part II

Party Organisation

Chapter Six

The Congress

The Role of the Congress

The European People's Party (EPP) Congress is the highlight of the party's activities. Held at least once every three years, in different venues, it is the forum for agreeing on the political line or programme, for deciding about changes to the statutes and for introducing the party's leadership and members to one another. The agenda is normally centred on the adoption of a general political manifesto, either pertaining to an upcoming election or dealing with a more specific policy area, and is sometimes accompanied by internal elections.

The Congress is made up of delegates from the member parties, from the recognised associations (for youth, students, women, workers, local and regional politicians, and small and medium enterprises), members of the EPP Group in the European Parliament (EP) who belong to the member parties, the EPP Presidency as a whole, national heads of party and government, and European Commissioners who belong to a member party. The number of delegates is weighted according to the EPP's share of Members of the European Parliament (MEPs), individual delegates being elected by the appropriate body in the member party. Member parties set their own rules for delegate selection.

The Congress follows rules of procedure that are decided on and occasionally amended by the Political Assembly (formerly, the Political Bureau). Until the 1986 Congress in The Hague, the content of Congress documents was finalised in advance by a consensus at the executive levels. This was done to take account of minority opinions, and with an eye to maintaining unity. During the Congress, delegates could discuss the documents, propose footnotes and deliver commentaries, but they were not allowed to make any changes to the documents themselves. Since the 1988 Luxembourg Congress, all changes or additions to Congress documents have been made by majority vote of that body. This includes changes and additions to a Congress document prepared by a programme commission, but amendments must be proposed before an announced deadline.

The Congress meets in plenary and also in a number of working groups, usually two. Both the plenary and the working groups are permitted to make decisions. The working groups are intended to reflect the composition of the Congress, and the delegations' votes count for half the number of votes they hold in the Congress plenary. Proposals which in the working groups are supported by fewer than a third of delegates present will be rejected; those supported by a two-thirds majority will be accepted. Those that fail to achieve a two-thirds majority, but are approved by more than a third of the delegates present, will – after a position has been recommended by the programme petitions committee – be presented to the plenary of the Congress for decision. A final decision is taken by simple majority.

A new quality of cooperation and integration emerged when this voting procedure, made possible by a growing policy consensus and, in particular, the increased ability of delegates to understand each other's points of view, was put into place at the end of the 1980s. It should be kept in mind that the principal problem for international or supranational understanding in a federation of like-minded parties is not that individual national elements hold differing interests or ideas and quarrel. The main problem lies, rather, in the different languages, cultures and constitutional backgrounds of the various countries from which partners and colleagues come. They bring with them different ways of speaking and behaving. Styles of debate and rhetoric vary from country to country. To those who are inexperienced, everything seems alien, making understanding, let alone trust, more difficult. The skill of the interpreters does not change this. The use of interpreters enables people to understand each other, but – with more languages and language combinations leading to longer translation processes – it also reduces spontaneity. In this way, compared with a national party congress, a European party congress is less lively. The problem is very difficult to overcome.

Since its founding in 1976, the EPP has organised almost 20 Congresses, most of which have not been held in Brussels. Some took place in the run-up to EP elections; others have had a longer time frame and dealt with long-term developments within the party or in certain policy areas. The titles of the documents that were presented, discussed, amended and finally accepted reflect the diversity of the party's development and orientation.

Historical Congresses

Table 3. EPP Congresses (1978–2010)

Number	Place, Date	Document Title
I	Brussels, 6–7 March 1978	Together towards a Europe for Free People – Political Programme
II	Brussels, 22–23 February 1979	Striving Together for a Europe of Free Citizens – Election Manifesto for the First Direct Elections to the European Parliament
III	Cologne, 1–2 September 1980	The Christian Democrats in the Eighties: Securing Freedom and Peace, Completing Europe
IV	Paris, 6–8 December 1982	Establishing Peace – Preserving Freedom – Uniting Europe
V	Rome, 2–4 April 1984	Action Programme for the Second Parliamentary Term of the EP, 1984–1989
VI	The Hague, 10–12 April 1986	EPP's Tenth Anniversary – Striving for a People's Europe
VII	Luxembourg, 7–8 November 1988	On the People's Side – Action Programme for the Third Parliamentary Term of the EP, 1989–1994
VIII	Dublin, 15–16 November 1990	For a Federal Constitution for the European Union
IX	Athens, 12–14 November 1992	Basic Programme
X	Brussels, 8–10 December 1993	Europe 2000: Unity in Diversity – Action Programme for the Fourth Parliamentary Term of the EP, 1994–1999

Table 3. (continued)

Number	Place, Date	Document Title
XI	Madrid, 5–7 November 1995	EPP – Force of the Union
XII	Toulouse, 9–11 November 1997	We Are All Part of One World
XIII	Brussels, 4–6 February 1999	On the Way to the 21st Century – Action Programme for the Fifth Parliamentary Term of the EP, 1999–2004
XIV	Berlin, 11–13 January 2001	A Union of Values – Basic Document
XV	Estoril, 17–18 October 2002	A Constitution for a Strong Europe
XVI	Brussels, 4–5 February 2004	The EPP: Your Majority in Europe – Action Programme for the Sixth Parliamentary Term of the EP 2004–2009
XVII	Rome, 30–31 March 2006	For a Europe of the Citizens: Priorities for a Better Future
XVIII	Warsaw, 29–30 April 2009	Strong for the People – Action Programme for the Seventh Parliamentary Term of the EP 2009–2014
XIX	Bonn, 9–10 December 2009	The Social Market Economy in a Globalised World

Brussels 1978 and 1979

The first EPP Congress took place on 6–7 March 1978 in Brussels. It was a Congress that was exclusively dedicated to the founding of the party.[148] The EPP's second Congress took place on 22–23 February 1979, again in Brussels, less than one year after the first. Its central task was to prepare for the imminent first direct elections to the EP, and above all to decide on

---

[148] See chapter two of the present volume and the annexes.

the electoral platform. The party's positions had been firmed up and large parts of the 1978 Political Programme updated. The Congress approved a text that had been proposed by rapporteur Hans August Lücker, voted through by the programme committee chaired by Wilfried Martens and agreed to by the party's executive bodies.[149] The following individual chapters of the electoral platform were presented to Congress: 'A Europe of Freedom and Solidarity', 'A Responsible Europe', 'A Democratic Europe Open to the Outside [World]', 'A Europe of Economic and Social Progress and Full Employment' and 'A Europe of Freedom, Security and Peace'.[150]

Cologne 1980

The third Congress took place in Cologne on 1–2 September 1980, a full year after the first direct EP elections. The Congress document was entitled 'The Christian Democrats in the Eighties: Securing Freedom and Peace, Completing Europe'. The main speakers were Flaminio Piccoli, the Political Secretary of the Italian Christian Democrats (Democrazia Cristiana, DC), and Jean Lecanuet, President of the French Democratic and Social Centre (Centre des démocrates sociaux, CDS). Their subjects were East–West relations, the energy crisis, the international currency and financial issues, and the further development of the European Community (EC) into a political union. Agreeing on the principal party lines, but with different emphases, they analysed the problems and marked out the political framework within which Christian Democrats would have to face the challenges of the next decade.

This Congress took place in Germany, not least because of the approaching Bundestag elections on 5 October. In that sense it was an EPP contribution to the election campaign of the German Christian Democrats (Christlich Demokratische Union, CDU) and the Bavarian Christian Social Union (Christlich-Soziale Union, CSU). The candidate for the position of chancellor, Franz-Josef Strauss, used the opportunity to present his programme to an international public. He appealed to Europe's Christian De-

---

[149] The text is reprinted in Eva-Rose Karnofski, *Parteienbünde vor der Europa-Wahl 1979. Integration durch gemeinsame Wahlaussagen?* (Bonn, 1982), 219ff.

[150] The text of these reports appeared in *CD-Europa* 2 (1979), published by the Christian Democratic Group in the European Parliament, which reports on the second EPP Congress.

mocrats, and particularly to those with government responsibility in their countries

to accelerate the realisation of European Union despite all the well-known difficulties. Thus Europe, as a strong second pillar of the Atlantic Community of values within the NATO alliance, can be true to its historic calling and to its role on the world political stage. On that too depends the free and peaceful future of Europe and of its peoples.[151]

With the greater responsibilities it now bore as a result of the direct elections to the EP, the EPP, like the other parties and parliamentary groups, needed to develop clear, concrete policies within the EC. This meant that the Congress, for all the agreement on basic issues, pointed out political differences too. The consideration of measures to overcome unemployment was referred to the Political Bureau, for instance, since no agreement could be reached.[152] Much of the debate concerned the future of the EPP itself: organisational and financial problems were handled in the same way as questions of internal coherence, and the party's profile in terms of its political programme.

Paris 1982

The fourth Congress took place in Paris from 6 to 8 December 1982, under the title 'Establishing Peace, Preserving Freedom, Uniting Europe'. Several programme texts were discussed and agreed on, including two on foreign political subjects, 'Freedom in Peace and Security' and 'Development Policy: A Duty of Solidarity and a Vital European Responsibility', and two on social policy, 'Internal Peace: Economic and Social Policy' and the 'Action Programme for Small and Medium-sized Businesses in the European Community', accompanied by a resolution on the institutional strategy of the EPP.[153]

---

[151] Hans August Lücker and Karl Josef Hahn, *Christliche Demokraten bauen Europa* (Bonn, 1987), 159.

[152] See Rudolf Hrbek, 'Die europäischen Parteienzusammenschlüsse', in Werner Weidenfeld and Wolfgang Wessels (eds.), *Jahrbuch der Europäischen Integration 1980* (Bonn, 1981), 261.

[153] The texts are printed in *Beiträge zur europäischen Wirtschafts and Sozialpolitik. Programme, Berichte, Resolutionen 1981–1984*, vol. 4, EPP Documentation series (Brussels, n.d.), as well as *Freiheit, Gerechtigkeit, Friede. Die Verantwortung der Europäer in der Welt, Programme, Berichte, Resolutionen*

For the first time, CDU Chair Helmut Kohl took part in an EPP Congress as Chancellor, having won this office for the first time a few weeks before. On 1 January 1983, Kohl's new federal government would take its turn as president of the Council of the EC. Kohl used his appearance at the EPP Congress to signal initiatives that would advance the EC on the road to a European Union (EU): 'If we fail to make the decisive step forward during this decade, we will have wasted a historic opportunity; we must not delay European unity in the hope that better times and conditions will arrive and allow us to start all over again'.[154]

Paolo Barbi had succeeded Egon Klepsch as Chair of the EPP Group at the beginning of 1982. Both critical and constructive, his fiery speech to Congress called for a better-organised EPP and an expanded Secretariat that could be set to work continuously. In particular, he said, the EPP Secretariat must be in a position, on the basis of precise knowledge of the problems and of mainstream opinion in the member parties, to take responsibility for identifying common positions and drafting practical proposals.[155]

Rome 1984

With imminent direct elections to the EP in June 1984, the fifth Congress, held in Rome on 2–4 April 1984 was mainly concerned with agreeing on and presenting a new action programme. A proposal by Arie Oostlander, director of the Scientific Institute of the CDA, had been elaborated by a programme committee chaired by Secretary General Thomas Jansen. After exhaustive discussion, the Political Bureau unanimously approved the text. Neither the text nor proposed amendments were to be discussed. Indeed, the whole point of the Congress was to have the Action Programme for the next legislature endorsed by acclamation. Speakers also had to limit themselves to commenting on and publicising the policies set out in the programme.

This process demonstrated not only an anxiety about being seen to agree. It also showed an aversion to open, democratic debate – in other

---

1981–1984, vol. 5, EPP Documentation series (Brussels, n.d.). See also the report about the Paris Congress in *CD-Europa* 1 (1983).

[154] Lücker and Hahn, *Christliche Demokraten*, 175.

[155] See Paolo Barbi, *Napoli–Strasburgo e ritorno. I cinque anni al Parlamento Europeo* (Napoli, 1985), 151.

words, a lack of mutual trust. The unanimity rule flagrantly contradicted the EPP's claim to be a party. However, in the end it did not matter much. Public consciousness, and therefore the electorate, was far from being sensitised enough to take offence. And the competitors, the Social Democrats and the Liberals, had not yet got any further along this road; on the contrary, they had little more to offer than general declarations. The EPP Action Programme was designed to be comprehensive, omitting no important issue, and was also very progressive in its statements on European policy.

The four chapters of the Action Programme for the second legislature of the EP dealt with the principal tasks and goals of European policy. They were entitled 'More Jobs and Employment in Europe', 'For Human Development in a Viable Europe', 'For a Stable Peace Based on Justice and Solidarity' and 'More Democracy and Unity in Europe'. An introductory chapter dealt with the basic convictions that informed the programme, under the headings 'Opting for Europe', 'More Jobs through Economic Recovery', 'Possibilities for the Citizen', 'Caring for the Environment and Nature' and 'A Just Peace'.[156]

The detailed Action Programme once again reflected serious efforts to deal with a series of existing problems. Proposals had to be as specific as possible. So apart from analytical observations and statements of principle, every chapter also contained lists of demands that were binding on the party. Compared with the 1978 Political Programme, conceived more as a basic programme than an action programme, the 1984 text evinced a new kind of consensus. It reflected systematic and continuous cooperation within the EPP structures, notably in the EPP Group during the previous legislature. The Action Programme's closing sentences expressed a new confidence, a result of the experience of unity and mutual action: 'The EPP occupies a key position in the European Parliament, with the political power of substantially influencing the decision-making process . . . The EPP is the party which steadfastly, determinedly, and unanimously stands for the creation of a United States of Europe'.[157]

The 1984 Congress was held in what, for these purposes, was a thoroughly unsuitable hotel on the outskirts of Rome. Proceedings suffered not only from public indifference but also from the fact that the leadership of the Italian host party seemed neither politically engaged nor overly in-

---

[156] Text in vol. 6, EPP Documentation series (Brussels, n.d.).
[157] Ibid.

terested in organising the event. Ciriaco de Mita, Political Secretary of the DC, carried the final responsibility. Throughout his entire period of office, from 1982 to 1990, he showed not even a mild interest in cooperation within the framework of the EPP.

But the Congress was successful as far as its content and the participation of delegations were concerned – with the exception of the Italians, who mainly distinguished themselves by their absence. The Congress helped significantly in giving member parties a political direction and in shaping the EPP's internal cohesion. Crucial in this respect were the committed speeches of EPP President Leo Tindemans and government leaders Kohl (Germany), Ruud Lubbers (the Netherlands), Martens (Belgium), Jacques Santer (Luxembourg) and Garret FitzGerald (Ireland).

However, a political–cultural demonstration at the Rome opera house, designed to open the European election campaign, was a fiasco. Those responsible in the DC leadership had neglected to invite potential participants, namely Christian Democrats and sympathisers from Rome and the surrounding area. About four hundred delegates and Congress guests, along with cultural and political stars, were marooned in the huge hall, which could hold two thousand people. However, the Italians were able to compensate for such organisational deficiencies through their gastronomic contributions; eventually there was the kind of atmosphere in which friendship thrives.[158]

The Hague 1986

The sixth Congress took place during EPP President Piet Bukman's term of office; he was also Chair of the CDA. It was obvious that this Congress, scheduled to run from 10 to 12 April 1986, should be held in The Hague. This time it was not a matter of preparing for a European election but of presenting the growing movement. The process of European integration had been palpably enlivened by Jacques Delors's initiative for a single European market, called 'Objective 1992', and by Spain and Portugal's joining the EC in January 1986. The EPP had admitted the Christian Democratic parties in both countries as members: the Spanish Democratic Popular Party (Partido Demócrata Popular, PDP), the Catalonian Democratic Union (Unió Democràtica de Catalunya, UDC) and the Basque

---

[158] See Barbi, *Napoli–Strasburgo e ritorno,* 281, which indicates that this disaster can be traced to an intrigue between De Mita's *correnti* and Andreotti.

National Party (Partido Nacionalista Vasco, PNV), and the Portuguese Democratic and Social Centre (Centro Democrático e Social, CDSp).

The organisational and material conditions at the congress centre in The Hague were excellent. The Dutch host party had ensured, with the help of social events under their aegis on the Congress margins, that the atmosphere was equally agreeable. This was also the first EPP Congress in which a number of representatives of non-European parties from the Christian Democrat International (CDI) took part, notably from Latin America, among them CDI President Andres Zaldivar (Chile) and the President of the Christian Democratic Organisation of America (Organización Demócrata Cristiana de América, ODCA), Ricardo Arias Calderon (Panama), as well as its Secretary General, Aristides Calvani (Venezuela).

The 1986 Congress focused on two large subject areas: economic development and environmental problems. Two comprehensive Congress documents had been prepared and were used as the basis of discussion; with amendments, they were voted through. Two working groups, chaired by Fernand Herman and Hanja Maij Weggen, respectively, carried out intensive debates. The rapporteurs were Elmar Brok and Reinhard Klein. A whole raft of proposals for resolution were discussed, some of them controversial. Overall, this Congress allowed for much more debate than previous Congresses had: there were not just speeches, but also spontaneous contributions. As at all previous Congresses, EPP government leaders were once more the stars of the show. Lubbers, Martens, Santer, FitzGerald and Kohl all spoke at a concluding demonstration to which, apart from the delegates, CDA members from far and wide had been invited. Only de Mita was conspicuous by his absence.

Luxembourg 1988

Another European election, in June 1989, was imminent at the time of the seventh Congress, and the meeting was conceived as a prelude to the consciousness-raising and mobilisation campaign being undertaken ahead of the election campaign proper. The Congress was held on 7–8 November 1988 in Luxembourg, with Santer as Chair. It was used to discuss and vote on the Action Programme for the 1989–94 legislature. The Programme Committee, chaired by Lutz Stavenhagen, Minister of State in the German foreign ministry, had prepared a very detailed document. The Congress was also presented with numerous proposals for amendments and additions.[159]

In contrast to the 1984 Congress in Rome, this Congress became a forum for debate and decision making. The individual chapters of the Action Programme were debated in four parallel study groups. Decisions were reached with a two-thirds majority on proposals from member parties, the EPP Group and associations. Under standing orders, proposals which were neither accepted nor rejected by the study groups were put to the vote in the plenary session, where a simple majority decided their fate. Since the study groups had been given the right to make decisions for the whole Congress, their composition had to mirror that of the Congress. This meant that each group had to contain a quarter of all the members of individual delegations with the right to vote. This procedure ensured lively, sometimes heated, debate. It involved not just the classic issues of practical European policy, but also socio-political problems such as the permissibility or otherwise of artificial insemination. The answers to such questions challenged fundamental positions held by member parties.

The 1988 Action Programme's chapter headings contained the essential message that Christian Democrats in the EPP wanted to send to the voters in EC Member States: 'We Are Strengthening Democracy in the Community and Creating the European Union', 'We Are Creating Europe as a Modern, Effective, and Humane Economic and Social Space' and 'We Are Shaping a Humane Society in a Europe of Citizens'.[160] The Action Programme was passed. This was the first time in the history of a European party that a programme was agreed on by a majority after open debates, as was an election manifesto containing, in abbreviated form, the main EPP demands and a call to the citizens of Europe to support the EPP in realising them.

Alongside the logo introduced when the party was founded, the green 'E' with a circle of stars, the Luxembourg Congress presented a new logo: a blue heart containing a circle of 12 stars. This was to illustrate the Congress slogan 'EPP – the heart of Europe'. Since then, this has been the logo

---

[159] See reports and documentation in *EPP Bulletin: Communiqués from the General Secretariat* 5 (June/July 1986).

[160] The text appears in *An der Seite der Bürger. European People's Party Plan of Action 1989–1994* with reports by Santer, Klepsch, Stavenhagen and Thomas Jansen, as well as other documents from the seventh EPP Congress in Luxembourg (Melle, 1989).

of the EPP Group in the EP. A public event, a 'Euro show' with folkloric entertainment and popular music, marked the conclusion of Congress, graced with the presence of EPP heads of government. About two thousand people took part, most having travelled from Belgium, the Netherlands, France and Germany. They were joined by a good number of Italian, Greek, Spanish and Portuguese party supporters living and working in these countries.

Dublin 1990

The importance of the eighth Congress, held in Dublin on 15–16 November 1990, was the new perspective created for the EC and for all of Europe by the political watershed of 1989. The Central and Eastern European countries had freed themselves of Communism and Soviet hegemony; the Berlin Wall had fallen and the reunification of Germany had become a reality. The foundations were laid for a new and decisive move towards a European Union, one that had to be equipped to take in the newly liberated countries.

A meeting of the European Council had been fixed for 15 December. On the basis of the vision for a new EU, the Council was to call for an Intergovernmental Conference (IGC) to revise the Community treaties. The EPP's ideas on this were to be discussed at the Dublin Congress. A draft document written by Paul Dabin had been worked on by a committee chaired by EPP Vice-President Alan Dukes, then Chair of the Family of the Irish (Fine Gael, FG). The draft contained not only a description of the form of the desired EP and its institutions but also suggestions on how to reach that goal.

This was the first Congress chaired by Martens, then still Prime Minister of Belgium, and President of the EPP from May 1990 onwards. The Congress was held under the rules tried out in Luxembourg two years before. Two working groups discussed and passed the two chapters presented by the Programme Committee along with amendments. The result was published under the title 'For a Federal Constitution for the European Union'.[161]

In the weeks and months that followed, this document would have a considerable influence. It set the direction not only for the EPP Group but

---

[161] The text is printed in *EPP Bulletin: Communiqués from the General Secretariat* 5/6 (December 1990).

also for the majority in the EP. Indeed, there was a consensus among the majority parliamentary groups along the lines of this document at the so-called Assizes, an informal meeting of MEPs and national parliamentarians held in Rome on 30 November. Christian Democratic heads of government and ministers involved in negotiating the new treaty largely followed the ideas developed in the EPP's paper.

While the Congress was already overshadowed by the leadership crisis in FG, the simultaneous leadership crisis in the British Conservative Party attracted the Congress's particular attention. The impending resignation of Margaret Thatcher reactivated expectations of a closer association between the British Conservatives and the EPP. Several Conservative politicians were present at the Congress as observers. The British press followed the discussions and the result particularly closely. In addition, this was the first-ever EPP Congress attended by delegates invited from Central European countries. The Slovene Prime Minister and leader of his country's Christian Democrats, Alojze Peterle, addressed the Congress, a fact that clearly articulated the new dimensions of European politics. Dublin was also the first EPP Congress of José María Aznar, Chair of the Spanish People's Party (Partido Popular, PP). The PP had become an EPP observer a couple of weeks before the start of the Congress.

Athens 1992

The effect of these changes on traditional policymaking and indeed on the whole style of Christian Democratic vocabulary was palpable. A new consensus had to be forged. This was secured by adopting a new Basic Programme at the ninth EPP Congress on 12–14 November 1992 in Athens. The Basic Programme laid down the political direction the EPP has followed since: a solid commitment to continuing towards a united Europe, which in turn meant institutional deepening and geographical enlargement of the Union on the basis of the social market economy and a political ethic guided by the Christian image of man.

Congresses had by then been held in all EC Member States with EPP member parties – except for Greece. It was now its turn, and the Greek New Democracy (Néa Dēmokratía, ND) was keen to host the event. After the summer 1990 elections in Greece, the party, under Konstantin Mitsotakis's leadership, was once more in government. The legendary hospitality of the EPP's Greek member party was a good augury for a splendidly organised Congress and an appropriate venue.

Discussion of the new Basic Programme was the main focus. Member parties and the EPP Group in the EP, as well as the recognised associations, had proposed a large number of amendments and additions. These were discussed in working groups and in the Congress plenary session. The vote, as had become routine since the Luxembourg Congress in 1988, was by simple majority. Obviously, an attempt was made to secure broad agreement on controversial issues.

The British and Danish Conservative MEPs had been admitted to the EPP Group in the EP in May 1992, and the intention was to bring the Scandinavian Conservatives into the EPP. Therefore, it was important for the EPP leadership to restate unambiguously both the party's Christian Democratic identity and its commitment to a federal Europe. There were also important tactical reasons for this. There had been resistance on the part of a number of member parties, notably the Dutch, Belgians and Italians, to widening the EPP's membership. Agreement had only been reached by a commitment to deepen it as well.

Tensions caused by different attitudes and cultures took centre stage in Athens, and were also evident in the debates. This was because of both the strategic and tactical implications mentioned above and the exceptionally ambitious and extensive text of the draft programme. However, the work was eventually finished, and the various protagonists could see that their fears of being brought down by their partners or pushed along the wrong track were mainly groundless. Tensions abated, giving way to general agreement about what had been achieved.[162]

Brussels 1993

The Athens Congress had instructed the Political Assembly, then Bureau, to set up a committee to prepare a draft Action Programme based on the results of the Congress. Martens was once again the Chair, with almost the same members as those in the Basic Programme Committee. Work began in the spring of 1993. The draft had to be completed by the summer if the text was to be debated and voted on at the next EPP Congress, scheduled for December 1993.

An 'action programme' in the EPP tradition is not merely an electoral platform containing a few hard-hitting demands. It contains detailed poli-

---

[162] See the second annex of this volume.

cies for all key problem areas, and is in effect the party's and the parliamentary groups' 'breviary' for the next parliamentary session. But, of course, it must be available in time for the electoral campaign, so that candidates and campaigning parties can refer to it. And it must also be ready in time for member parties to include its contents in their own campaign preparations. Even more important, they must be able to take it into account in their own policy statements. The rapporteurs for individual chapters of the 1993 Action Programme were Klaus Welle from the CDU, Jacques Mallet from the French Democratic and Social Centre (Centre des démocrates sociaux, CDS), France), Herman from the Belgian French-speaking Christian Social Party (Parti social chrétien, PSC), Ferruccio Pisoni from the DC and José-Maria Gil-Robles from the PP, along with President Martens and Secretary General Thomas Jansen.

At the tenth EPP Congress, held in Brussels on 8–10 December 1993, working groups and the plenary session debated amendments to the Programme Committee's draft text. They eventually decided the final version by majority vote; that procedure was by now tried and tested. Changes and additions had, as before, come from member parties, the associations and the parliamentary Group. The dates of the EPP Congress coincided with the meeting of the European Council at the end of the Belgian presidency (9–10 December 1993). The Congress had been preceded by a Summit of party and government leaders, where Jean-Luc Dehaene, Lubbers, Santer and Kohl spoke at the first meeting. Because of the Summit, Brussels was full of journalists from every leading country, but neither this opening session nor the Congress itself made any noticeable impact in the press. Attention was concentrated on national affairs. Journalists or their editors had clearly not yet registered that a congress held by a European party in this context, or indeed any context, might be relevant. Reporters were also distracted by the appearance of Salman Rushdie at the European Socialist EP Group, meeting at the same time in the Espace Léopold in Brussels. This, too, gives an insight as to how opinion-makers saw European parties.

Madrid 1995

The EPP held its eleventh Congress in Madrid on 5–7 November 1995. The thematic slogan was 'EPP – Force of the Union'. On the surface, this Congress was almost entirely dominated by the expected victory of the PP in the election that had been announced for the following March. The Congress opened with a public rally in the Real Madrid football stadium, with about 5,000 participants. Aznar, the PP's unchallenged leader and Vice-President of the EPP, was able to show the Spanish public that his party belonged to a strong and trustworthy political family in Europe. What was more, Aznar had a significant, indeed leading, role in this family.

The political content of the Congress was dominated by the position the EPP had to take at the 1996 IGC that prepared the Treaty of Amsterdam. The basis for discussion was a draft position paper entitled 'Ability to Act, Democracy, and Transparency – the European Union on the Road to a United Europe'. This had been developed during the course of 1995 by a working group chaired by the German MEP and EPP Group Vice-President Hans-Gert Pöttering, who was later to become Group Chair and President of the EP.

The paper set out and explained the arguments for the EPP's classic policies for a democratic, federal EU, along with proposals for overcoming the shortcomings of existing procedures. The Madrid Congress saw parties from Sweden, Denmark, Finland and Norway take part for the first time. They had been accepted as members in March of the same year. Given the increased number of member parties, and the fact that the enlargement process was still incomplete, the party leadership proposed changes to the statutes to allow for flexible reaction to the political challenges of the time. More than the other European parties, the EPP has always devoted a great deal of energy to developing structures for forming opinion and for decision-making. It wanted to be able to adapt to the demands of European politics, and so to the development of the Union's political system, as well as to strengthen the EPP's own effectiveness as a political body.

Revision of the statutes at the Madrid Congress was aimed above all at strengthening the party leadership by increasing the power of the Presidency, which in the future, from 1999 onwards, would be elected not by the Assembly (then Bureau) but by the Congress. The Presidency was also reduced in size by cutting the number of vice-presidents from thirteen to six. In advance of European Council meetings, the EPP Presidency was to meet with heads of government from EPP parties in a 'mini-Summit'. The Conference of Party and Government Leaders was to meet only when key decisions had to be made about the party itself. A new body, the EPP Council, was to meet regularly to decide on policy. The Council was to be composed of leaders of member parties, or their secretary generals, and the EPP Presidency. The Bureau of member party delegates would meet more often and continue to advise on day-to-day issues, taking any necessary decisions.

Toulouse 1997

The twelfth Congress, held in Toulouse on 9–11 November, was above all about policy: how to deal with the effects of globalisation, unemployment and issues connected with EU enlargement. Employment policy was especially topical; only a few days later the European Council was to hold an extraordinary (first) meeting on this issue. Employment was therefore a central question for the 'mini-Summit' held in the margin of the Congress. On the basis of the understanding the leaders were able to reach, Congress agreed on a strategy paper that was also to set the direction to be taken by the European Council.

At last, an agreement on another controversial matter was reached, setting the right direction for the European Council. The issue was EU enlargement, which was on the agenda for the Luxembourg Summit on 8 and 9 December. One of the key questions was whether all enlargement candidates were to start negotiating membership immediately or would negotiations begin as the European Commission had suggested, with those who were best qualified to join the EU in the near future. The EPP Congress resolution was based on a compromise formulated by Pöttering. It avoided discriminating against individual candidates, but still allowed for the differentiation necessary in the negotiations. This was the line finally adopted by heads of state and government.

A comprehensive programme paper entitled 'We Are All Part of One World' was debated at the Toulouse Congress. The document had been the subject of discussion and consultation for more than a year, and it had received hundreds of amendments. Divergent points of view and areas of mutual incomprehension had been building up since new member parties were admitted in 1995. That was now obvious to everyone. The parties from the Nordic countries, especially, were notable for taking positions that seemed barely reconcilable with those of the classic EPP parties. It amounted to a direct confrontation between a Liberal orientation and the Christian–Social position. Nonetheless, it proved possible to build a sturdy consensus, and, after substantial modification, to secure unanimous support for the comprehensive programme paper. That this could be achieved is evidence of the willingness and capacity to integrate the political forces brought together in the EPP.

Brussels 1999

The circumstances surrounding the thirteenth Congress, held in Brussels on 4–6 February 1999, were very particular. The euro had been introduced only a few weeks before, a qualitative leap in the history of European integration that marked a huge achievement in European construction, and a process largely shaped by EPP parties. At the same time, member parties had lost a large number of national elections in the previous two years, and the EPP had to accept that this would probably mean a marked loss of influence. What was more, the 1999 European elections were approaching. Congress had to debate and reach agreement on the 1999–2004 Action Programme, which would be the basis of the EPP campaign.

The party made it clear in this campaign that it felt sure it could win enough votes to become the largest European parliamentary group. This would to some extent balance its losses at national level. Another significant matter on the agenda was the integration of the European Union of Christian Democrats (EUCD) into the EPP, which in practice meant a further concentration of moderate parties in the EPP institutions. In short, the eventful Brussels Congress had a great deal on its plate.[163]

At the Congress it was noted that EPP values had spread into the Socialist and Social Democratic camps. There were in Europe, said President Martens, 'Socialist parties without Socialism', more or less copies of Christian Democratic parties, which had jettisoned their political ballast. Armed with the mystical notion of the 'Third Way', they were now raiding the EPP's constituency base throughout Europe. Since no one has a copyright on ideas, the EPP had to fight to prevent Socialist domination in Europe.

The big package of reforms called 'Agenda 2000' was central to the policy debate about further European integration. Its object was to create the preconditions for successful EU enlargement. The EPP position was clear: the reform of the EU institutions, its finances, the Common Agricultural Policy and the structural funds had priority over enlargement.

---

[163] Klaus Welle, 'Reform of the European People's Party 1995–1999', in Hans-Joachim Veen (ed.), *Christlich-demokratische und konservative Parteien in Westeuropa 5* (Paderborn-Munich, Vienna, Zurich, 2000), 541–66. See p. 562.

The party was committed to a European Constitution which would set out unambiguous rules, based on the subsidiarity principle, about who would do what at European, national, regional and local levels. Regional, political and social differences had to be defended against any form of centralisation, protected and encouraged in all their multiplicity. In the area of institutional reform, the EPP called for more decision making by majority vote, a reweighting of votes in the Council based on relative population size and reform of the Commission.

The core EPP economic goals were to decrease state intervention and create a more market-friendly macroeconomic environment. The proposal to coordinate economic policy at Community level had the broad agreement of delegates. But employment policy came under the provisions of subsidiarity: fighting unemployment was essentially a matter for the Member States and the regions. No one disputed that social security systems needed a radical shake-up if they were to fulfil their real function, which was to provide a safety net for the genuinely needy. There was a parallel emphasis on the responsibility of individual citizens. This shift of paradigms was not only a matter of financial *force majeure* but also an expression of the Christian image of man to which EPP member parties are committed.

The Congress marked an important change of direction for the party's organisation. For the first time, it was the Congress which elected the Presidency. Martens was elected EPP President for the third time, and the following Vice-Presidents were elected: Margareta af Ugglas, John Bruton, José-Manuel Durão Barroso, Karl Lamers, Kostas Karamanlis and Wim van Velzen. Following a change in the statutes designed to ensure there was a better balance, the number of vice-presidents was increased to seven, and the Bulgarian Foreign Minister Nadezhda Mihaylova was elected Vice-President, too, becoming the first representative of the Central and Eastern European member parties in the EPP Presidency. Ingo Friedrich was also elected to the Presidency as Treasurer; Pöttering, Chair of the EPP-ED parliamentary Group, and the Secretary General were ex officio Presidency members.

Welle, Secretary General since 1994, had become Secretary General of the EPP-ED Group, and was succeeded by Alejandro Longo Agag, a former personal advisor to Aznar. Another change in the statutes turned the Conference of Party and Government Leaders into the EPP Summit. Those with a right to attend now included the leaders of the main opposition parties in those EU Member States where there was no EPP member party in the government.

Berlin 2001

The fourteenth EPP Congress took place in Berlin on 11–13 January 2001, the German EPP parties being the hosts for the first time since the 1980 Cologne Congress. CDU Chair Angela Merkel and CSU leader Edmund Stoiber welcomed a number of Prime Ministers: Aznar of Spain, Jean-Claude Juncker of Luxembourg, the Austrian Chancellor Wolfgang Schüssel, the Slovak Mikuláš Dzurinda, Victor Orban of Hungary, Edward Fenech Adami of Malta and Mart Laar from Estonia.

But the line-up of heads of government was thinner than in the past. Representing their parties in Berlin, Silvio Berlusconi, Bruton, Barroso and Karamanlis were all either out of office, or had not yet won an election. The 1999 European elections had been held not long before, and the 2004 European elections were still a distant prospect. For the EPP it was an opportunity to discuss and agree on a new basic programme, though it was not formally to be called a programme but a 'document'. The Athens Programme, in other words, was to remain valid. However, the word 'document' was an understatement. The Berlin Basic Document took the Athens Programme forward and renewed it, though without supplanting it. While referring to common values, it looked for the right way to deal concretely with the challenges facing Europe in a time of overwhelming change.

Apart from the Congress document, entitled 'A Union of Values', resolutions were adopted on a wide variety of subjects, reflecting the range of issues that concerned the parties and associations drafting them. There was a veritable kaleidoscope of issues, as there is in the Congress document itself: EU enlargement, social development, cooperation between international organisations, the Cyprus question, Mediterranean policy, illegal arms trafficking, terrorism, euthanasia, health risks, BSE (bovine spongiform encephalopathy, or 'mad cow disease') and the post-Nice process.

Next to the Basic Document, the 'Berlin Declaration' proposed by the EPP-ED Group in the EP and accepted by Congress, was especially topical, coming only a few weeks after the December 2000 'Declaration on the Future of Europe' by the European Council in Nice. The Berlin Declaration set out the main points of the EPP's position in the debate about the

Constitution. The Congress document 'A Union of Values' made a further demand, again after a proposal by the EPP-ED Group, that 'this Constitution, which must be democratically adopted by the peoples [of Europe], should be elaborated by a body similar to that used to draft the Charter on Fundamental Rights'.[164]

Estoril 2002

The fifteenth EPP Congress was held in Estoril, close to Lisbon, from 17 to 18 October 2002. Barroso had been Prime Minister of Portugal since his party, the Social Democratic Party (Partido Social Democratico, PSD), had won the parliamentary elections in April of the same year. He was able to welcome about 1,000 delegates and guests from all over Europe. The Political Bureau met on the eve of the Congress opening and agreed to accept the Estonian Pro Patria Union (Isamaaliit) party as an associate member, and to accord observer status to the Polish Civic Platform (Platforma Obywatelska, PO) and the candidate parties from Croatia and Albania. The EPP was now represented in every country of Europe. The participants at Estoril again included several heads of government: Berlusconi from Italy, Aznar from Spain, Juncker from Luxembourg, Jean-Pierre Raffarin from France, Schüssel from Austria, Fenech Adami from Malta, Kjell Magne Bondevik from Norway and Dzurinda from Slovakia. Because of a government crisis, Jan Peter Balkenende, Dutch Prime Minister only since May 2002, was at the last minute unable to attend.

The Estoril Congress was dominated by debate on the European Constitution, which had been going on in the European Convention since 28 February under Valéry Giscard d'Estaing as Chair. But until then the Convention had essentially confined itself to searching for a general consensus on the character and identity of the Union, and its future development and shape. The real work on the text of the Constitution was to start in the autumn. The Congress document 'A Constitution for a Strong Europe', elaborated by a committee jointly chaired by Martens and Wolfgang Schäuble, made its appearance at just the right moment. Furthermore, it was based on important preparatory work and resolutions, not the least relevant of which were the decisions that had been reached at the Berlin Congress. The EPP and its parliamentary Group had been working on

---

[164] *EPP Yearbook 2000/2001*, CD-ROM (EPP Secretariat). See also the third annex of this volume.

these issues both ahead of the Convention and once it had begun. The Berlin Declaration's basic demands are reflected in the chapter headings of 'A Constitution for a Strong Europe': 'Lived Subsidiarity and Solidarity: Division of Competencies between the European Union and Member States', 'Integration of the Charter of Fundamental Rights in the Constitution', 'Reform of the EU Institutions', 'New Ways of Achieving Greater Democracy', 'More Transparency and Closeness to the Citizens' and 'Our Responsibility: Reforms Now!'.

The document's 53 points set out precise suggestions and demands for the text of the European Constitution. They represented a compromise between doctrinaire European federalism and what would win consensus support, not just in the Convention but also in the enlarged and now much more experienced EPP. This compromise did no harm at all to the coherence of the EPP's strategy in the Convention or to the force of the EPP's arguments and proposals to the Convention. Indeed it was the EPP, as declared in the introduction with justifiable pride, which was 'the first European party to put forward a comprehensive concept for a European Constitution. For the Christian Democrats, and for the centrist and reformist parties in the EPP, the heirs of Europe's founding fathers, this is both a challenge and a duty.'[165] The solid and well-considered ideas articulated in this document were a key precondition for the unity of the representatives at the EPP Convention Group, led by MEP Brok, and they became a driving force in the body charged with writing a European Constitution.

Apart from the Congress document, several other resolutions were agreed on, including the process of EU enlargement, the fight against terrorism, illegal immigration, sustained development policy, human trafficking and the transatlantic relationship. Martens was elected EPP President for the fourth time, receiving the votes of 88% of Congress delegates. Statute changes increased the number of Vice-Presidents from 7 to 10. The following were elected: Antonio Tajani (Italy), Bruton (Ireland), Alberto João Jardim (Portugal), Peter Hintze (Germany), Karamanlis (Greece), Mihaylova (Bulgaria), Orban (Hungary), van Velzen (Netherlands), Pierre Lequiller (France) and Bo Lundgren (Sweden). Friedrich was re-elected as Treasurer. Antonio López-Istúriz succeeded Alejandro Agag as Secretary

---

[165] 'A Constitution for a Strong Europe', document adopted by the EPP Congress, Estoril, 18 October 2002 (www.epp.eu).

General.[166] Sauli Niinistö, President of the European Democratic Union (EDU) since 1988, was named the second Honorary President of the EPP along with Tindemans, its founding President. The integration of the EDU and the EPP was now complete.

Brussels 2004

The EPP's sixteenth Congress was held on 4–5 February 2005. For the fifth time, the venue was Brussels. The main objective was to reach unanimity on the electoral programme and prepare for the European elections of the same year, so the 2004–2009 Action Programme for the next legislative period was central to the debate. A secondary objective was to raise the party's profile at an extraordinarily significant time in the EU's history. The IGC, which was concerned with the draft constitutional treaty developed by the European Convention in 2003, was still searching for a compromise on the issue – controversial until the last minute – of voting procedures in the Council. There was no real doubt that there would be a positive result, however, and that this would soon be concluded. The formal entry of the 10 new Member States, due to take place on 1 May, was imminent.

The 2004–2009 Action Programme is divided into six chapters dealing with the following themes: a dynamic, competitive economy committed to solidarity and creating jobs; answering citizens' growing need for security; sustainable development; European youth policy; Europe and its place in the world, and a strong and well-governed Europe. A detailed introduction first describes the EU as an ambitious, challenging project, still incomplete and full of promise, and reminds readers of the fact that the EPP has always solidly supported and promoted this project. After that follows, with reference to the emergence and the role of the EPP in the history of European integration, a summary of the fundamental positions the EPP has taken, and which were set out in the Basic Programme in Athens 1992. The central role of human dignity is stressed, along with the significance of cultural and ethical values. These same values inform the realisation of the European project, not only in terms of how the necessities of life

---

[166] Following his marriage to a daughter of Spanish Prime Minister and PP leader José María Aznar, Agag decided to resign his position in the EPP to avoid, from the outset, any potential conflict between family and political loyalties.

related to living together are to be practically organised, but also in terms of meaning.

Apart from the Action Programme and the manifesto, the Brussels Congress once again adopted resolutions on current or specific problems raised by individual delegations. These included resolutions concerning developments in Croatia, Belarus, Cyprus, Moldova and Romania, and on terrorism, security of life for older people and the Lisbon Agenda. A resolution under the title 'Capacity to Act – Democracy – Transparency' addressed itself directly to the IGC with a call to complete its work on the draft European Constitution before 1 May, thus respecting the Convention draft and enhancing rather than diluting the effectiveness of the Union after enlargement.

Another resolution concerned the 'Appointment of the Future President of the European Commission'. This proposal subsequently turned out to be fairly explosive and eventually became the determining factor in the procedure by which Barroso was appointed. The Convention's opinion, set out in the draft Constitution, was that the European Council should have to take into account the results of the European elections when proposing a candidate. On this basis, the EPP called on the Council Presidency not to make its proposals known before the European elections, so that the results could be respected. Heads of state and government were expected to respect the spirit of this provision in the Constitution. The EPP also called on EPP heads of state and government to take the necessary initiatives to reach consensus on a joint candidate in the relevant party and parliamentary group committees. Finally, the EPP-ED Group was called on to not support any candidate who did not come from the political family that had won the most votes in the European elections.

Rome 2006

The IGC finally agreed on a Constitutional Treaty but with the 2005 referenda in France and in the Netherlands turning it down, the new treaty was politically dead. A diagnosis made by many pointed at a lack of connection between the general public and the so-called political elite. This was not the first time the problem had been raised. What was different from previous occasions is that it had now affected two founding members of the EU, two countries that incidentally were ruled by EPP member parties in that particular period. It was largely felt in the EPP that an answer to these problems had to be made. Practically speaking, a document was prepared in which the EPP took a clear – and positive – stance on EU enlargement (despite enlargement fatigue) and on institutional reform (despite the negative referenda), particularly on the European Constitution as the final outcome of the Convention on the Future of Europe.

The statement explicitly referred to both the diagnosis and the remedy:

> It is essential to identify the problems and shortcomings of our integration model. We want a Europe of citizens. Therefore, we want to strengthen the principle of subsidiarity, solidarity and personalism, in the framework of a rational division of competences. We also want a transparent institutional structure, able to organise a network of political activity, with the simplification of decision-making procedures and clarification of spheres of responsibility. A network that is able to stimulate the political contribution of the citizens. A European Union of 25 plus Member States presents new challenges as to how to respond to the concerns of its citizens.[167]

The document entitled 'For a Europe of the Citizens – Priorities for a Better Future' was adopted at the seventeenth EPP Congress on 30–31 March 2006 in Rome. Forward Italy (Forza Italia, FI) under the leadership of its President, Prime Minister Berlusconi, acted as host. Besides the document and a long list of resolutions on different topics, the Congress saw the re-election of President Martens and Secretary General López-Istúriz. Members of the Congress also elected 10 new Vice-Presidents: Hintze (Germany), Michel Barnier (France), Orban (Hungary), Mario David (Portugal), Tajani (Italy), Camiel Eurlings (The Netherlands), Jyrki Katainen (Finland), Enda Kenny (Ireland), Jacek Saryusz-Wolski (Poland) and Peterle (Slovenia), along with Friedrich as Treasurer. An impressive number of party and government leaders addressed the Congress. During the Congress, the thirtieth anniversary of the party was celebrated. At this occasion, a new party logo was introduced.

Warsaw 2009

For the first time, an EPP congress was organised in one of the new EU Member States from Central and Eastern Europe. The city of Warsaw welcomed 2,000 people on 29–30 April 2009 for the eighteenth EPP Congress. Unlike all previous electoral congresses, the Congress was organised as late as six weeks prior to the 2009 EP elections. The venue for the kick-

---

[167] 'For A Europe of the Citizens – Priorities for a Better Future', document adopted by the EPP Congress, Rome, 31 March 2006 (http://www.eppcongress2006.org).

off of the EPP electoral campaign was not without symbolism. Twenty years after the collapse of the Eastern bloc and five years after the 2004 enlargement, the EPP rally took place in the capital of the largest of the new states, Poland. During the Congress, Poland's role in these developments was regularly referred to. In order to emphasise the triumph over Communism, the Congress took place in the Palace of Culture and Science (where the last congress of the Polish Communist Party had taken place). A speech by special guest Lech Wałęsa, the former President of Poland, reminded every Congress delegate of this particular 'European' history.

Most of the attention went to the adoption of the electoral manifesto entitled 'Strong for the People'. It consisted of five chapters: 'Creating Prosperity for Everyone', 'Making Europe a Safer Place', 'Combating Climate Change', 'Tackling the Demographic Challenge' and 'Uniting Europe on the World Stage'. A shortened version was meant to play a role during the election campaign. Traditionally, a number of resolutions were adopted but this time, not surprisingly, the emergency resolution on the 'Global Financial and Economic Crisis' was the most important one. The Congress also presented Barroso as the EPP's candidate for a second term as President of the European Commission. A large number of party and government leaders explicitly supported his candidacy by attending the Congress and addressing its delegates. The EPP showed confidence in its electoral performance. Although the British Conservative Party and the Czech Civic Democratic Party (Občanská demokratická strana, ODS) were officially still part of the EPP-ED Group, the prospect of having a parliamentary group without them already made things easier. It was foreseen that after the elections, synergies between the party and the Group would increase.

Bonn 2009

Less than eight months after Warsaw, the EPP held its nineteenth Congress on 9–10 December in Bonn. After the installation of the second federal government led by Chancellor Merkel, and in the run-up to the Landtag elections in North Rhine-Westphalia, the CDU together with the CSU welcomed more than 1,500 delegates to the premises of the former German Bundestag. Twenty years after the fall of the Berlin Wall, this site was historical and largely symbolic, especially since the Congress was dedicated to the concept of a *Soziale Marktwirtschaft*. The Congress document 'The Social Market Economy in a Globalised World' applied this concept to the new reality of a European economy recovering from the most severe crisis since the Second World War, while at the same time facing increasing challenges from new economic powers such as China and India. In the document, the social market economy model is presented as the way out of the crisis:

> We strongly believe that such a renewed Social Market Economy as an open system can become the trademark of our economic and social success in Europe and worldwide. Today we must find ways to make our economic structures more efficient and our financial systems more sound and robust within an adequate regulatory and supervisory framework. We must revisit our existing economic and social structures by fostering processes of innovation, by embedding our European economies in a wider international context and by making them more open to social mobility, economic dynamism and inclusion through more innovation and entrepreneurship . . . Only by applying these policies will we be able to maintain our lead on the global path towards growth and prosperity for all citizens.[168]

Next to the issue of economic recovery, a great deal of attention was devoted to the resolution on climate change. Climate change had already been on the EPP's agenda, at a so-called EPP Convention in Madrid in early 2008. Within the framework of the 2009 United Nations Conference on Climate Change, COP15, in Copenhagen, the EPP 'urged the EU and its Member States to remain united and steadfast in our effort to reach legally binding commitments on emission reductions as soon as possible to ensure that the "two-degree target" is reached in order to avoid dangerous climate change'.[169]

Furthermore, the Bonn Congress saw the modification of the party's statutes and internal regulations – among other things, the Political Bureau was renamed the Political Assembly – and Martens and López-Istúriz were re-elected as President and Secretary General, respectively. As well, Friedrich, as Treasurer, and 10 Vice-Presidents were elected for another three-year term: Hintze (Germany), Barnier (France), Orban (Hungary), David (Portugal), Tajani (Italy), Corien Wortmann-Cool (the Netherlands), Katainen (Finland), Kenny (Ireland), Saryusz-Wolski (Poland) and Rumiana Jeleva (Bulgaria). In addition to a long list of party and government

---

[168] 'The Social Market Economy in a Globalised World', document adopted by the EPP Congress, Bonn, 9–10 December 2009 (www.epp.eu).

[169] 'The UN Climate Change Negotiations in Copenhagen', resolution adopted by the EPP Congress, Bonn, 9–10 December 2009 (www.epp.eu).

leaders who were EPP members, all major EU institutions were represented by EPP politicians: Barroso (European Commission), Herman Van Rompuy (European Council) and Jerzy Buzek (EP). Thanks to the enactment of the long awaited Treaty of Lisbon, just a few days before the start of the Congress, the EPP was clearly at the helm of a Europe poised to become more democratic, transparent and efficient towards its citizens, as well as the larger world.

Chapter Seven

The Political Assembly and the Presidency

The daily business of the European People's Party (EPP) is not handled by the Congress, which meets only every two years or so, but by the Political Assembly and the Presidency. The EPP statutes describe the Political Assembly, formerly known as the Political Bureau, as 'the strategic organ' of the party, while the Presidency – often referred to by its Latin name, Presidium – is 'the executive organ'.[170] These two compose, so to speak, the parliament and the government of the EPP. Both party bodies are chaired by the EPP President.

The Political Assembly

The Political Assembly consists of delegates sent by national parties that are members or associate members of the EPP, the number of delegates being proportional to the number of members each party has elected to the European Parliament (EP). The Assembly also consists of a number of ex officio members: the members of the EPP Presidency; the EPP Group Presidency; members and leaders of national delegations; Chairs of member parties and member associations; insofar as they belong to member parties, the European Commissioners and the President and/or Vice-Presidents of the EP; members of the Presidency of the Committee of the Regions; and Presidents of the EPP Groups in the Committee of the Regions and in the parliamentary assemblies of the Council of Europe (CoE),

---

[170] EPP Statutes approved by the EPP Congress on 10 December 2009 in Bonn, articles 15 and 11, respectively.

the Western European Union (WEU), the Organisation for Security and Cooperation in Europe (OSCE) and NATO.[171]

The political task of the Political Assembly is to ensure the unity of the party as an active body and to influence European politics in the spirit of the party programme and the policies set by Congress. In addition, the Assembly has a number of governance tasks: it adopts the annual accounts and the budget, decides on the amount of the annual dues to be paid by the members and elects the Deputy Secretary Generals on the recommendation of the Presidency. Another important role is that of gatekeeper, charged with approving new members, recognising member associations, excluding members and revoking the recognition of member associations. Finally, it is also the Political Assembly's duty to convene the Congress and decide its agenda and regulations.[172]

These tasks make the Political Assembly the most important decision-making body, especially as far as intra-party affairs are concerned. It is the forum for agreement or argument between the party leadership, the member parties' representatives and the EPP Group in the EP. The goal of the parties representatives is, as a rule, to formulate for the Group leadership the political line the Group should take. The Group leadership's aim is to bring the Political Assembly over to the Group's way of thinking. Insofar as these conflicts are about issues requiring expertise in European Union (EU) politics and experience in dealing with EU institutions, the Group representatives usually have the advantage. By contrast, in general political and organisational matters or problems related to membership applications, the member parties tend to be better at getting their way.

The Political Assembly meets at least four times a year and, in practice, at least every two months. Most meetings take place in Brussels, in one of the rooms allocated to the EPP Group in the EP buildings. Votes and election procedures are decided by absolute majority of the members present. Majority voting has been a common practice since the EPP's founding in 1976. Ideally, the Political Assembly and, particularly, the EPP President who chairs it, seek to establish general agreement. At the same time, the Assembly strives to respect members' individual identities and not to make

---

[171] Ibid. All the members have voting rights. A number of other people are also members of the Political Assembly but without voting rights, for instance, the secretary generals of the above-mentioned organisations. See article 15.

[172] EPP Internal Regulations approved by the EPP Political Bureau on 9 December 2009 in Bonn, article Ib.

unrealistic demands of any delegation. Enormous attention has to be devoted to the member parties' sensitivities.

Over the years it has become possible to consolidate the EPP's internal relations – integration and consensus have grown – as member party delegates have shown more and more willingness to back down on an issue in order to facilitate reaching a common position. However, experience has shown that a 'national' position, in particular a bottom-line position, is frequently made the subject of a stubborn rearguard action for home consumption only. Sometimes the temptation to adopt a purely national position is very strong. It is, after all, not always simple to defend a decision 'at home' which breaks with a much-loved national way of seeing things or which conflicts with a national habit, even when this seems the sensible and necessary course from a larger European perspective.

The Presidency

In the original constitution, the Presidency was accorded a purely ceremonial role. In the new 1990 statutes, however, it became an independent body responsible for guaranteeing the EPP's political presence at all times, ensuring the Political Assembly's decisions are carried out and overseeing the work of the Secretary General, especially in budgetary matters. This puts the Presidency at the head of the party, a forum in which the President, Secretary General and Treasurer discuss internal, operational and administrative proposals and initiatives with the Vice-Presidents, before these concerns reach the Political Assembly. To enhance this continuity, the Presidency meets at least eight times a year. In practice, meetings often take place before the Political Assembly or the EPP Summits.

Since 1990 the election of Presidency members has taken place every three years; previously, it was every two. Until 1999 the Presidency members were elected by the Political Assembly, then Bureau. Since then the President, 10 Vice-Presidents (until 1993, the number was 6), the Secretary General and the Treasurer are elected by the Congress. A number of other people are ex officio members of the Presidency: the EPP President, who chairs the Presidency; the Secretary General; the Chair of the EPP Group in the EP; and, to the extent that these people are affiliated with the EPP, the President of the European Commission, the President of the European Council, the High Representative of the Union for Foreign Affairs and Security Policy and the President of the EP.[173] Not surprisingly, May 1990 saw the first of a long series of contested elections for Vice-Presidents, simply because there were more candidates than places. This was a clear indication of the Presidency's enhanced powers. But it also gave expression to a new openness and the triumph of democratic procedures over diplomacy.

The President

The office of President of the EPP is not an official body of the party, such as the Congress, the Political Assembly or the Summit, but it is clear that the position is central to the party's functioning and has been crucial in its development. Formally, the President represents the party both internally and externally, and leads and chairs the Congress, the Political Assembly, the Presidency meetings and the Summits. Since 1999 the President has been directly elected by the Congress; before 1999 the election took place in the Political Assembly, which was then named the Political Bureau. The President is elected for a three-year term. The success of the EPP depends to a large extent on the President's political commitment and the success of his or her initiatives. This is certainly true of the relations with the various member parties and the relevant circles in European politics. The President must lead the party – by means of integration and moderation – while at the same time ensuring that it is present and recognised as an important and strong political organisation in EU politics.

Coincidently or not, every single EPP President so far has been a political figure from the Benelux countries: the Belgians Leo Tindemans (1976–85) and Wilfried Martens (since 1990), Piet Bukman (1985–87) from the Netherlands and Jacques Santer (1987–90) from Luxembourg. These centrally located countries on the fault line between the dominant European cultures are destined to play the part of balancing the interests of their larger neighbours. Politicians from the Benelux are evidently thought to have

---

[173] EPP Statutes approved by the EPP Congress on 10 December 2009 in Bonn, article 11. Until the merger of the EUCD with the EPP in 1999, the EUCD Presidents were ex officio members of the Presidency as well: Kai-Uwe von Hassel (1973–81), Diogo Freitas do Amaral (1981–83), Giulio Andreotti (1983–85), Emilio Colombo (1985–92), Wilfried Martens (1993–96) and Wim van Velzen (1996–99).

the greatest talents in directing such a multi-voiced choir. Moreover, this is no mere Christian Democratic prejudice – the Socialists and Liberals have predominantly recruited their party chairs from the Benelux countries as well. And there are practical reasons for this. Politicians from these countries master a number of languages and, beyond their linguistic skills, have a great capacity for appreciating different political cultures. Apart from all that, a President who lives in Brussels itself, or in Luxembourg or The Hague, is better able to follow events centred on Brussels and regularly take part in the necessary meetings and discussions.

Three of the four EPP Presidents were prime ministers of their countries at the time of their election. Only Bukman held no government office, being 'just' a party chair when he became President of the EPP. Choosing personalities who had already made a name for themselves in Europe as prime ministers reflected the EPP's need to make itself known. Right from the start it had to secure a political profile. It is an obvious advantage for the party to have a President who can count on the kind of public attention that a recognised figure in high office attracts, a trend that was later to be followed by the Socialists and the Liberals.

The tactic of choosing a government leader for party President also has its inherent disadvantages. The priorities of a national government leader are determined by his or her administration's political agenda. A head of state who is at the same time head of a European party will constantly face the temptation to use this position in the service of his or her work as a head of government. Even the President who has no intention of using the European party as an instrument of national government still cannot let his or her hands be tied by the national party position. This dilemma, however, has never raised serious problems or conflicts, although it is ever-present. One way in which it has shown itself is in the priority given by such leaders to the activities and appointments of their own governments. In addition, the expectations of EPP bodies, and the policy outlines they have laid down, have been at odds with the reserved, cautious, diplomatic way of speaking that has characterised all EPP Presidents who were simultaneously either prime ministers or members of a government. As seen below, there were also differences between them resulting from their different temperaments.

Leo Tindemans, 1976–85

Tindemans was the EPP's first President. He had been known as 'Mister Europe' ever since writing the now famous 1975 report commissioned by heads of European Community (EC) states and governments entitled 'Report on European Union'. Tindemans was extremely popular everywhere in Europe, not only because of his political position or role on the national and international stage, but also because of his charismatic, communicative nature. He was able to express himself with ease in numerous languages and was a highly effective communicator via the media. Tindemans already had a long history of distinguished government service. He was essential to his party, the Flemish Christian People's Party (Christelijke Volkspartij, CVP), as an electoral campaigner. He was equally important as a linchpin in the great debate between the Flemish and the Walloons about a new political order in Belgium, a dispute that threatened to rip the country apart. His character and inclination also made him indispensable to the party's international relations. From 1965 until 1972, Tindemans played an important role in multinational party cooperation as Secretary General of the European Union of Christian Democrats (EUCD), acquiring pertinent experience and knowledge.

Tindemans had not been involved in founding the EPP because of his duties as Prime Minister of Belgium, but because his past career and political profile identified him with the new party, the proposal to elect him the first President met with spontaneous general agreement.[174] When in 1978 Tindemans had to give up his position as Prime Minister, he became Chair of his party, the CVP. The following June, the first direct elections to the EP saw him elected in Flanders. The size of his victory broke all records and has not been surpassed: he won almost one million preferential votes, one third of the electorate. However, his hopes of succeeding Simone Veil as President of the EP in 1982 were dashed when the Chair of the EPP Group, Egon Klepsch, threw his hat into the ring, having obtained the nomination from the EPP Group in autumn 1981. Eventually the Dutch Socialist Piet Dankert was elected.

Before the end of the year, Tindemans had left the EP to become Foreign Minister of Belgium in Prime Minister Martens's government. During his last years as EPP President, up to April 1985, Tindemans was simultaneously a foreign minister, a post he held until mid-1989. It became clear

---

[174] Wilfried Martens, *Europe: I Struggle, I Overcome* (Brussels, 2009), 41–2.

that being a foreign minister was not a satisfactory 'moonlighting' job for the president of a European party. The hands of a foreign minister are bound by the exigencies of diplomacy, far more so than those of a prime minister. If a foreign minister wishes to successfully represent his national government, he cannot – as the EPP party bodies expect – publicly make critical demands or launch radical initiatives. The conflict of interest became too much, and after the second direct elections to the EP, Tindemans gave up his position as EPP President.

Piet Bukman, 1985–87

Bukman was elected Tindemans's successor in March 1985. His background was in farmers' associations, and he had been Secretary General and President of the Dutch Christian Farmers and Gardeners Association (Christelijke Land- en Tuinbouwbond). His political career had begun in the Anti-Revolutionary Party (Anti-Revolutionaire Partij, ARP), where he had been part of the executive and involved in the party's management during the 1970s.

When he was elected President of the EPP, Bukman was party Chair of the Dutch Christian Democratic Appeal (Christen Democratisch Appèl, CDA), which he had led since 1980. Indeed, he was the first Chair of the party, which had been created by merging three Christian confessional parties – the Catholic People's Party (Katholieke Volkspartij, KVP), Christian Historical Union (Christelijk-Historische Unie, CHU) and ARP – which had previously existed side by side. He proved himself a capable, significant figure in guiding the integration process. Moreover, Bukman already had a good name in the EPP as well: he had been a member of the Presidency since 1982, and had been, though this was never formalised, the party's first Vice-President. When Tindemans was unable to attend meetings of the Political Bureau because of his commitments as foreign minister, Bukman always chaired them. His style was characterised by a certain discipline and determination to get results.

His experience as Chair of a party that had grown out of a number of different components made Bukman very aware of the integration problems that the EPP had to solve. He showed great understanding in these matters, and took an active interest in the practical and organisational questions of transnational cooperation. His door was always open.

Bukman faced the disadvantage of not being well known internationally, since he had never held high government office. He compensated for that by being always available, making time to travel for the EPP and speaking throughout Europe. Bukman's successful term came to an end after the two-year mandate, when he was made Minister of Development Aid in Ruud Lubbers's government after the autumn 1986 election. The political conventions in Holland prevented Bukman from holding both a government office and a party position. A special dispensation from the prime minister, confirmed by the cabinet, was necessary for Bukman, as a new government minister, to complete the last few months of his mandate as President of the EPP.

Jacques Santer, 1987–90

Since the EPP had now had both Belgian and Dutch Presidents, it was obvious to those looking for a successor to Bukman that the next party President should be a Luxembourger. Santer was an obvious candidate. He had been the Prime Minister of the Grand Duchy of Luxembourg since 1984. Prior to that, he had led the Christian-Social People's Party (Chrëschtlech-Sozial Vollekspartei, CSV). Both as Secretary General (1972–74) and as President (1974–82), he had been present for all the important decisions relating to the EUCD and the founding of the EPP. For a few years, from 1975 to 1979, Santer had been a deputy in the EP, for a time even a Vice-President of the EP. He was familiar with party work at all levels, knew the European scene from the perspective of an experienced parliamentarian and had served in a number of governments.

Santer's presidency lasted from March 1987 until May 1990. Prolonging his mandate from two to three years anticipated a change made to the party's statutes in autumn 1990. The practical political reason for prolonging the mandate was that in spring 1989, when elections for a new EPP Presidency were due, the European election campaign was already in full swing. It was therefore decided to postpone the election of the Presidency until the line-up of the next parliament was known, in particular, the members of the EPP Group.

As President, Santer held many of the qualities people had valued in Bukman. He was always present, he was willing to engage himself, he was interested in solving practical problems and he was aware of (sometimes contradictory) expectations. Santer had certain advantages over his predecessor, arising from his position as a head of government. Doors were always open to him, and there was no obstacle to his moving in the highest political circles. Also, membership in the European Council, becoming increasingly important for the development of what was then the EC, now the EU, was an incalculable advantage for the EPP President. Santer knew how to develop that advantage and how to use it.

Wilfried Martens, 1990–

In May 1990 Santer passed the baton of leadership to Martens, his colleague in the European Council, who had been Prime Minister of Belgium since 1979. Martens, more than anyone else, seemed predestined to be President of the EPP. He had been one of the EPP's founders; along with Hans August Lücker, he had proposed the party's first constitution and its first political programme. At the time, he was Chair of his national party, the Flemish CVP. As a head of government and member of the European Council for more than 10 years, Martens had gained a deep understanding of European politics.

His period of office falls into five distinct phases, the first lasting up to the moment he resigned as Prime Minister of Belgium (from May 1990 to February 1992), the second until his election as Chair of the EPP Group in the EP in July 1994 and the third spanning the period when he was President of both the party and the EPP parliamentary Group, from 1994 to 1999; the fourth phase of Martens's leadership stretches from 1999 to 2004 and the fifth from 2004 to the present.[175]

The first period, leading up to the Treaty of Maastricht, proved to be essential for European politics and the development of European parties. The building of a political Europe put the President of the EPP – and the EPP as a whole – under much greater pressure than had been experienced under previous Presidents. Martens also took very seriously the resulting need for the party to be active in Central and Eastern Europe. Above all there was considerably more need for discussion between member parties, and especially, between EPP heads of government. In this connection, too, Martens's dual function was very useful.

The second phase of Martens's party leadership was characterised by his being free of the burden of government office, and able to devote himself

---

[175] For an extensive review, see Ibid.

entirely to his job as EPP President. It was especially fortunate that Martens was available full time to the EPP between February 1992 and July 1994. He brought to his work as head of the EPP the wealth of experience gained in office, along with the elaborate network he had cultivated. In the job of party leader, he had become familiar with the EPP's structures and all the relevant problems of European politics. His very concrete engagement was, not surprisingly, linked with his ambition as a top-ranking politician to look for more and more important challenges. His only public platform during this period was as President of the EPP; he held no other position apart from his membership in the Belgian Senate. Martens's path to new and more important positions was tied to the success of the EPP and that, in turn, was a great advantage to the party.[176]

Martens's leadership of both the party and the Group characterises the third phase. This dual role held the possibility of breaking down or eliminating the contradictions and tensions that inevitably existed between the party and the parliamentary Group. It offered the chance, as well, of improving coordination between the two organisations, their bodies and secretariats. Indeed, Martens's dual leadership was bound up with the hope that it would enable the party and parliamentary Group to harmonise their communication and cooperation structures. The resultant synergies, it was calculated, would improve the effectiveness of the EPP.

The fourth period of Martens's leadership was similar to the second. Undisrupted by other functions or commitments, he devoted himself exclusively to party business. He exploited these opportunities with his customary dedication and reliability. His political experience and the network of personal friends and acquaintances he had built up over the years were trump cards for the EPP. These factors, combined with Martens's good reputation as a solid and dependable partner, helped the party through a number of difficult situations and, above all, ensured the party could handle the great challenges posed by EU enlargement.

The fifth phase of Martens's presidency is characterised by his enduring commitment and party consolidation. Consolidation was made possible by the 2004 Regulation on the statute and the financing of European political

---

[176] Evidence of Martens's tireless involvement everywhere in Europe, along with what he was saying and doing, can be found in his *L'une et l'autre Europe, Discours européen 1990–94*, Brussels 1994.

parties and its 2007 amendment.[177] Through this new regulation, the EPP held and holds the necessary means to develop a true party organisation, including its own research foundation. These developments inevitably also colour the functioning of the party President, who can rely on a stable and transparently financed tool to propose policies and new ideas for Europe's future. In addition, the 1999, 2004 and 2009 victories of the EPP member parties in the European elections have delivered the people to bring these ideas to fruition. Even more, in part thanks to the role Martens played as party President, the EPP now holds an impressive number of important posts in the EU institutions and its Member States.

The Political Assembly and the Presidency are probably the party bodies least known outside the EPP. Nonetheless, they each fulfil an essential role in the functioning of the party that should not be underestimated. The Assembly and the Presidency are the places of 'those who are always there', politicians who regularly serve the EPP, be it as Members of the European Parliament (MEPs), delegated representatives of member parties or elected officials sitting on EPP bodies. Their commitment is essential when it comes to decisions and party activities. However, the most important and prominent role is that of the President of the EPP. This is the key position to examine in order to understand the party's development and its actual functioning.

---

[177] See chapter twelve of this volume.

Chapter Eight

The Summit

No other party body has seen a rise in importance and exposure equal to that of the European People's Party (EPP) Summit. The Conference of Party and Government Leaders, as the Summit was called until 1995, was not foreseen in the original formation of the EPP. However, in parallel with the crucial role in the integration of Europe that the European Council played in the 1980s and 1990s, the EPP started to organise meetings with prime ministers and leaders from member parties. After several years the meetings were formalised, and they now take place regularly. As a result of their success, they have become central to the functioning of the EPP. The most recent offspring of the EPP Summits are the Ministerial Meetings, which offer the opportunity for ministers of European Union (EU) Member States who belong to EPP member parties to coordinate policies in the run-up to Council of Ministers' meetings.

The Origins

The first attempt to organise an EPP Summit dates back to the early 1980s. In the spring of 1980, an initial meeting of EPP party leaders was called by EPP President Leo Tindemans and held in Strasbourg. The main topic was the possibility of strengthening the voice of the EPP in the decision-making processes of what was then the European Community (EC). The idea was that Christian Democrats in responsible positions at different levels should cooperate and coordinate their work. Tindemans argued that '[w]e should make efforts to establish a system for consulting each other on important national, European, and international problems'.[178] There was general agreement on the paramount importance of the EPP's leading figures regularly consulting with one another. It was therefore decided 'that there would be a meeting of the Bureau [now Political Assembly] at least three times a year in which all party leaders, our Prime Ministers and the most important ministers, would take part. These meetings of the Bureau should if possible take place before the convening of the European Council'.[179]

This decision was only put into practice in the autumn of 1983. However, the meetings took a different form than originally intended. Instead of meeting in the context of the regular sessions of the Political Assembly, party leaders met independent of them, with a separate agenda, and with only party and government leaders involved. Such a meeting took place on 3 October 1983 in the Chateau Val Duchesse in Brussels under Tindemans's leadership. There were three items on the agenda: the further political and institutional development of the EC; EPP tactics in anticipation of the 1984 European elections; and the cooperation of Christian Democratic parties in the EPP, European Union of Christian Democrats (EUCD) and Christian Democrat International (CDI).[180]

At the suggestion of Helmut Kohl, who had been Chancellor for only a year and was federal leader of the German Christian Democrats (Christlich Demokratische Union, CDU), it was decided to call another conference and also to invite heads of government belonging to member parties. Only a few weeks later, on 26 November, the first Cconference of Party and Government Leaders was held, once again in Brussels, but this time at the Palais d'Egmont. All the party leaders took part, along with all the Christian Democratic government leaders: Kohl (Germany), Wilfried Martens (Belgium), Ruud Lubbers (the Netherlands), Jacques Santer (Luxembourg) and Garret FitzGerald (Ireland). This meeting was the prelude to a series of such conferences, which for the time being had an informal character.[181]

---

[178] Description of the debate on 24 February 1980 in Strasbourg, EPP General Secretariat Archive.

[179] Ibid.

[180] Recommended agenda for the meeting of party leaders and the EPP Group, 3 October 1983, Chateau de Val Duchesse, Brussels, EPP General Secretariat Archive.

[181] Conferences of party and government leaders held before the 1990 Dublin Congress were as follows: 26 November 1983 in Brussels, 23 April 1985 in Luxem-

They were not provided for in the EPP statutes, nor were there any internal rules setting out either a procedure or purpose for them.

Why Hold These Summits?

It became a tradition that the first item on the agenda dealt with the development of the EPP, and with organisational and political problems that had arisen in the party's work. The debate would regularly be introduced by the President's report. These conferences of party and government leaders set a political direction that was of the greatest importance for the development of the EPP's structures, its sense of identity and its political strategy.

The second item on the agenda was, as a rule, EC politics; over the years the agenda of the upcoming meeting of the European Council came to dominate the discussion. Over time, the EPP Summit did in fact become a welcome opportunity to prepare for the official European Council meeting. Irrespective of the negotiating positions of governments, and especially of foreign ministries and the diplomatic corps, it was possible – among friends, so to speak, and not in the context of binding negotiations – to exchange ideas and test reactions to proposed solutions to problems. The Summit was also a forum for agreeing on tactics.

For party leaders in opposition at home, attending such meetings was not merely a chance to put across their own points of view. It was also their opportunity to inform themselves about positions and ideas to which they would not otherwise have access. These meetings also sometimes afforded an insight into the strengths and weaknesses of their own countries' negotiating positions, which was especially useful to members of the European Council. In any event, both government leaders and opposition leaders benefited from exchanges at the EPP Summit.

For a long time such meetings were prepared by means of an aide-memoire from the Secretary General that summarised the confirmed position of the EPP on agenda items and, in some cases, recommended a line to take. Only rarely would resolutions connected with meetings of party

---

bourg, 19–20 June 1985 in Rome, 9 November 1985 in Brussels, 1 March 1986 in The Hague, 30 May 1987 in Brussels, 30 May 1988 in Bonn, 19 October 1988 in Brussels, 17 February 1990 in Pisa and 21 October 1990 in Brussels.

and government leaders be made public. Especially when discussions were taking place ahead of European Council meetings, it was not in the interests of government leaders to be tied down; they needed to retain as free a hand as possible in their discussions with other government leaders. After 1988, when Santer took over as President of the EPP, it became the practice for the party leader to give his summary of the results of the conference to the press.

A Formalised Party Body

Only after the revisions to the 1976 original statutes, made at the Dublin Congress in 1990, did the Conference of Party and Government Leaders become a party body. At the Madrid Congress of 1995, the term 'EPP Summit' was officially introduced. Membership comprised the President and the Secretary General of the EPP; the Chair of the EPP Group in the European Parliament (EP); the President of the EUCD, where he or she belonged to a member party; Chairs of member parties (party leaders); the President of the EP, where he or she belonged to a member party and the President or a Vice-President of the European Commission, representing those who belong to member parties. The statutes, however, did not say much about the tasks of the conference, merely stating that the President should report to the Political Bureau, now the Political Assembly, on such meetings and the directions to be followed as a result of them.

To some degree, the importance of the conference corresponded first to the number of prime ministers who were affiliated with the EPP, and second, to the number of them who attended. During some periods, especially in the second half of the 1990s, the absolute number of EPP government leaders was indeed low, but attendance has never been a major problem, as it has been for the summits of the Socialists.[182] However, with the enlarge-

---

[182] Only one Christian Democratic government leader, namely the Italian Prime Minister Ciriaco De Mita (1985–89), took no part in the work of the EPP. As Political Secretary, he was also the leader of the Italian Christian Democracy (Democrazia Cristiana, DC) between 1983 and 1993. His attitude was in crass contrast to the European and international commitment of his predecessors (for instance, Mariano Rumor, Amintore Fanfani, Emilio Colombo, Arnaldo Forlani, Giulio Andreotti). This indifference was also symptomatic of the helter-skelter decline of his party during his period of office.

ment of the EPP, starting in the first half of the 1990s, the number of party leaders grew almost exponentially. This, of course, did not always help the efficiency of the meetings, especially when topics related to the European Council meetings were discussed. Kohl, in particular, favoured a more limited conference in terms of numbers and sometimes invited EPP government leaders to Bonn for informal meetings at the German Chancellery or at his holiday residence along the Rhine.[183]

Between 1996 and 1999, the EPP officially distinguished between (de facto large) Summits and so-called mini-Summits, limited to government leaders. Gradually, the rule that every member party was to be represented by either a government or opposition leader (and thereby prevent a party from being represented by both its government and party leader, given that in a lot of member parties these are two separate positions) helped to limit the number of participants. The same rule now applies to countries with more than one member party: only the largest party can attend the Summits (unless the member parties do not directly compete with each other, such as the CDU and the Bavarian Christian Social Union (Christlich-Soziale Union, CSU) or the Christian Democratic parties in Belgium). At the same time, Summit participation has been extended to representatives from junior government parties, if the senior government party is not a member of the EPP.

In addition to the number of people attending, the frequency of the meetings has also been on the rise. While in the early period, the Conference of Party and Government Leaders was convened only occasionally, now the EPP Summit takes place at least four times a year. This, of course, is directly related to the increased frequency of the European Council meetings. In conjunction with some informal meetings of EU heads of state and government, European Council meetings normally take place at the end of June and of December, plus one in spring and usually one in autumn. Since the Treaty of Nice, most Council meetings have taken place in Brussels, whereas before, the country holding the rotating Presidency chose the venue. This privilege now applies only to the informal European Council meetings. EPP Summits have also followed this trend, with most

---

[183] See, for instance, Wilfried Martens, *Europe: I Struggle, I Overcome* (Brussels, 2009), 142–4. This particular example of 24 March 1998, one of the last meetings with Chancellor Kohl, was also of particular importance for the development of the EPP.

now taking place in the Palais d'Egmont and the Palais des Académies in Brussels and the Boechout Castle in Meise, near Brussels.

Setting the Lines

Over the years, EPP Summits have acquired significance insofar as they have been a means to develop common positions and a common direction for meetings of the European Council. At the same time, the EPP's common position, to which was always added the voices of a number of natural allies, has in several cases been crucial to the European Council's capacity to make decisions. The EPP Summits have also been of primary importance with regard to intra-party matters. Officially, however, the Summits can only make recommendations. This policy reflects sensitivity to the role of the Political Assembly, whose composition is based on democratic and federal principles. The Assembly is not supposed to forgo its right to make decisions and only develop the practical consequences of what the 'chiefs' have decided, but it should be remarked that this sensitivity is of a somewhat theoretical nature. Practice has shown that the Assembly has more respect for the party leaders than vice versa.

The significance of the regular meetings of party and government leaders for the EPP's development as a European party is not to be found purely in the Summit's function as a leadership body. The other element is that such meetings have been the public manifestation of the EPP as a united, powerful organisation, one that has made a difference. The internal effect of the conference of EPP leaders has been to crystallise a feeling of integration among the leaders and to provide an opportunity to identify with the wider movement. One demonstration that these meetings have been a positive exercise is the fact that the European Social Democrats copied the example of the EPP and established a regular meeting of party leaders when they founded the Party of European Socialists (PES) at the start of the 1990s.[184]

---

[184] The PES statutes accepted at the first party Congress in The Hague, 9–10 November 1992, envisaged 'meetings of party leaders, to be called at least twice a year' and to which 'Social Democratic members of the European Community and the Council of Ministers' could be invited.

Before the tradition of holding Summits became established, leaders of member parties who were prime ministers limited themselves to making an appearance at the EPP Congress. This amounted as a rule to no more than an *acte de présence*. The role of party and government leaders has markedly changed since they began meeting regularly to discuss not only general political questions but also the development of the party. As members of a party body, they are directly involved in deciding the general directions the EPP takes and directly influence the process of building political will and strategic decisions. Some government leaders are also heads of member parties. This dual role gives particular weight to their input on all issues concerning the development of the party.

Kohl remains the longest-serving leader at these meetings to be simultaneously head of his party and of the government of his country. He was present from the beginning, and was from the outset the most influential participant. This was in no doubt because the CDU was also the most powerful party in the EPP, and the Federal Republic of Germany under Kohl's leadership progressively acquired prestige and therefore influence. But Kohl always had to offer the EPP more than just the weight of his country and his party. He always knew what he wanted. It was at his insistence that the first meeting of government and party leaders took place, in 1983. He almost always attended, rarely permitting himself to be represented by a substitute because of another engagement. As time passed he increasingly became the opinion leader in the EPP because of his exceptional experience as a party leader, and his national and international standing. Normally he was either the first to speak, or was asked to speak first, and what he had to say set the tone.

Once the Spanish People's Party (Partido Popular, PP) was admitted as a full member in the spring of 1991, its leader José Maria Aznar began playing an increasingly active role in the EPP. In the spring of 1993 he was elected Vice-President. With the rapid improvement in his party's position in the battle with Felipe Gonzales's Socialist Workers Party (Partido Socialista Obrero Español, PSOE), Aznar's own position grew stronger too. After the PP's great victory in the 1994 European elections, the party eventually took on the mantle previously held by the Italian Christian Democracy (Democrazia Cristiana, DC). Along with the German CDU, the PP was a major people's party from one of the larger European countries. Aznar exploited the responsibilities and the influence this entailed. During his period in office (1996–2004) he became one of the key member-party leaders in shaping the EPP's development, not in the least because of his presence and active involvement in the EPP Summits.[185]

Making a Difference

Overall, the EPP Summits have been especially concerned with issues related to intra-party matters (such as the acceptance of new member parties and the European parties statute), EU institutional reform (including the Convention on the Future of Europe) and the EU enlargement process. Participants do not, on such occasions, invariably agree with each or with every position taken by EPP member parties or the parliamentary Group. The situation at the national level, the sensitivities of their coalition partners, specific European political pressures – any one of these can mean that heads of government need to reserve the right to depart from the EPP line. But there is a clear value in discussing different points of view as seen from different areas of responsibility. To all those involved in the process of forming opinions and taking decisions, understanding the motives underlying different political positions is helpful. Over time, such meetings have contributed to building consensus, and in some cases, this intra-party consensus has been decisive for the position taken by the European Council.

One of the situations in which an EPP Summit proved to be crucial was the enlargement of the EPP Group to include the British and Danish Conservatives. Since the membership application was extremely sensitive and raised the question of the identity of the Group as well as the party, President Martens decided to refer the matter to the Conference of Party and Government Leaders. At a special meeting devoted solely to the issue, held at the Chateau Val Duchesse in Brussels on 13 April 1991, the EPP confirmed its Christian Democratic identity and its commitment to European federalism. But a signal was also sent that the EPP was ready to engage in closer cooperation with all popular parties committed to 'a similar social project, and with the same goals in terms of their European policy'.[186]

---

[185] Wilfried Martens, *Europe: I Struggle, I Overcome*, 115–16.
[186] Procès-verbaux, EPP Conference of Party and Government Leaders, Brussels, 13 April 1991, EPP General Secretariat Archive.

On this basis, the EPP Group was advised to undertake a process of 'orchestration' with Conservative party deputies to resolve various matters of both substance and practice. If this process proved satisfactory, the creation of a *Fraktionsgemeinschaft* – a parliamentary alliance, as found in the German Bundestag between the CDU and the CSU – was to be considered. The Conference of Party and Government Leaders studied the results of this on 14 February 1992 and concluded that they were sufficient 'for the *Fraktionsgemeinschaft* to be [put] into effect on May 1'.[187]

Another highlight of the meetings was connected with the part they played in 1991 in the origins of the Maastricht Treaty. There were three meetings of party and government leaders that year. On President Martens's initiative, Christian Democratic government leaders or their agents met twice that autumn – as a subcommittee of the Summit, so to speak – to ensure there was agreement on the new treaty.[188] The initiative did not come out of the blue. The backdrop to the meeting of 13 April, which focused on the issue of cooperation with the Conservatives, was already the 'perspective of a federal, democratic European Union and its later enlargement to the countries of Scandinavia and Central Europe . . . The evolution of the European Community into a federally structured Political Union, including a common currency and common security, is a high priority goal for the EPP'.[189]

The meeting of the Christian Democratic party and government leaders in Senninger Castle near Luxembourg was wholly devoted to preparing for the meeting of the European Council (Luxembourg, 28–29 June 1991). Six government leaders were present, all of whom would be at Maastricht: Kohl (Germany), Giulio Andreotti (Italy), Lubbers (the Netherlands), Konstantin Mitsotakis (Greece), Santer (Luxembourg) and Martens (Belgium).[190] The summary of what had been agreed upon read as follows:

---

[187] Procès-verbaux, EPP Conference of Party and Government Leaders, Brussels, 14 February 1992, EPP General Secretariat Archive.

[188] The first of these meetings took place in Brussels in September, the second on the periphery of the NATO Summit meeting on November 7–8 in Rome.

[189] Procès-verbaux, EPP Conference of Party and Government Leaders, Brussels, 14 February 1992, EPP General Secretariat Archive.

[190] Austrian Vice-Chancellor Josef Riegler was there for the first time, along with Maltese Prime Minister Edward Fenech Adami, since the Austrian People's Party (Österreichische Volkspartei, ÖVP) and the Maltese National Party (Partit Nazzjonalista, PN) had been accepted as associate members of the EPP.

The EPP explicitly supports the unity of the European Union and categorically rejects all proposals which amount to placing the Community, along with other forms of cooperation, under the intergovernmental tutelage of the European Council. The EPP therefore insists that the parallelism between Political Union and Economic and Currency Union is thoroughly justified; they are two different aspects of the same structure, namely the European Union.[191]

Just before the meeting of the European Council in Maastricht (9–10 December 1991), there was an additional EPP Summit in The Hague on 6 December, following which Martens could declare that:

We in the EPP are unanimously agreed that Maastricht must be a success. However, we are not ready to agree to a compromise which puts in question the democratic and federal development of the future Union, or the irreversible character of this development. In particular we emphasise our commitment to increasing participation and co-decision by the European Parliament over a wide jurisdiction and an extension of Community competencies in areas which are important to economic cohesion and the development of a social dimension.[192]

In agreement with Lubbers, who would chair the European Council in Maastricht, Martens had drawn up a list of the most important demands, which the EPP heads of government were unanimously committed to defending. The demands related to the following points

- the irreversibility of the timetable for establishing the Union and its democratisation,
- the indivisibility of the institutional structure,
- the investiture of the Commission and the extension of its mandate to five years,
- EP co-decision in legislation,
- the agreement (*avis conforme*) of the EP to new actions and in favour of changes in the treaty,
- the independence of the Central Bank and the EP's agreement to statute changes,
- the creation of an advisory Committee of the Regions,
- the strengthening of the statutes of the Court of Auditors,
- more majority voting in the Council,
- economic and social cohesion,

---

[191] Communiqué de presse, 21 June 1991, EPP General Secretariat Archive.
[192] EPP Conference of Party and Government Leaders, The Hague, 6 December 1991, EPP General Secretariat Archive.

- the number of German deputies in the EP, and
- recognition of the role of the European parties.[193]

A comparative analysis of the results of Maastricht European Council and the EPP demands shows that on only one point was there no progress at all, namely on the *avis conforme*. Seven of the demands were satisfactorily met. The achievements on four other points left open the hope of a breakthrough at the conference that was to revise the Maastricht Treaty, scheduled for 1996. In any event, it could be stated at the Conference of Party and Government Leaders on 14 February 1992 'that the results of Maastricht owe much to orchestration between Christian Democrats'.[194]

It was no coincidence that these two key EPP Summit meetings after 1983 occurred during the same period. Both concerned vital decisions that had to be made in two inextricably connected processes: the establishment of an EU, and the Europeanisation of the party system. To emphasise these highlights is certainly not to belittle the contributions of party and government leaders before and after that period. For example, the conference on 20 June 1985 in Rome, held a week before the European Council meeting in Milan (28–29 June 1985), was also important. Here it was decided to call an intergovernmental conference to effect a clear advance along the road to an EU. And without Christian Democratic support, Italian Prime Minister Bettino Craxi would scarcely have risked following Foreign Minister Andreotti's advice to put the proposal to hold an intergovernmental conference to the vote in Milan. The move was fiercely resisted by British Prime Minister Margaret Thatcher. The support of five Christian Democratic heads of government (Kohl, Lubbers, Martens, Santer, FitzGerald), and of Andreotti – who, with Craxi, was a key player because of the Italian Presidency of the European Council – was indispensable.

The conference in Pisa on 17 February 1990 was also of considerable historical importance. The following points were on the draft agenda: the reform and further development of EPP structures, the challenge resulting from the emerging democracies in Central and Eastern Europe, the consequences of the new situation for the European Community, and priorities

---

[193] Ibid.

[194] Analytical note for the Conference of Party and Government Leaders, 14 February 1992 (Pascal Fontaine, 21 January 1992), EPP General Secretariat Archive.

for 1990 and longer-term perspectives.[195] A report was prepared on the first point summarising the conclusions of a Working Group set up with the objective of strengthening EPP structures and 'Europeanising' member parties.[196] On the last point of the agenda, Chancellor Kohl was able during the EPP Summit to convince Andreotti of the need for German unification. Kohl used all of his persuasive powers to explain to his colleagues that German reunification did not pose any danger to the process of European integration, but on the contrary, offered numerous opportunities. Both processes, the European and the German, were bound up with each other and each affected the other.

Another and later highlight of EPP Summits was the controversy over recognising Turkey as an EU candidate and starting membership negotiations with it. This was on the agenda in 2002, with a favourable decision in December 2004. The range of sensitivities, attitudes and arguments expressed at the earlier EPP Summit made a substantial contribution to the eventual decision by the European Council, based on a compromise that had been proposed by a number of EPP heads of government. The nub of it was to clearly establish that starting negotiations did not necessarily mean that Turkey would eventually join the EU, and that there were other possibilities for the EU's long-term relationship with the country.

Another important decision by heads of state and government was prepared by and came out of the interplay between the EPP Summit and EPP-ED Group in the EP. This was the nomination of José Manuel Barroso as President of the European Commission in 2004. EPP-ED Group leader Hans-Gert Pöttering had earlier demanded that the heads of state and government not fix the candidate before the June European elections, basing his argument on a provision in the Draft Treaty establishing a Constitution for Europe. The EPP-ED parliamentary Group, he said, would not agree to any candidate who did not come from the political family that had won most support in these elections. The EPP Congress, held in Brussels in February 2004 adopted this measure as a resolution proposed by the EPP-ED Group. But after the European elections, the EU's Irish Presidency was not in a position to propose a candidate who would have the support of

---

[195] Aide-mémoire for the EPP Conference of Party and Government Leaders on 17 February 1990, EPP General Secretariat Archive.

[196] On the cooperation of Christian Democratic parties within the EPP, see the report of the Working Group on 'Reform', presented by Wim Van Velzen, EPP General Secretariat Archive.

both the heads of government and a majority of the EP. Luxembourg Prime Minister Jean-Claude Juncker was generally agreed to be the ideal candidate, but he ultimately rejected the position. An attempt was made by French President Jacques Chirac and German Chancellor Gerhard Schröder to circumvent the EU Presidency and pressure other heads of government into accepting Liberal Belgian Prime Minister Guy Verhofstadt. This proposal was decisively rejected by the EPP and failed.

Martens had been given the task by the EPP Summit of articulating an opinion on who from the ranks of the EPP might have a chance of being accepted by the European Council. He now brought Barroso into play, and it was Barroso who was finally nominated and also elected by the EP, despite obdurate resistance by the Socialists. The Socialist parliamentary Group found the result difficult to swallow, even though it was, after all, a success for the EP as a whole, and for the democratic principle of majority decision-making. In any event, following the election of Barroso, the Socialists and Greens continued the struggle by trying to undermine the candidature of several members of the Commission team picked by its new President. This was done in the parliamentary hearings to which the candidates had to submit themselves. In the case of Rocco Buttiglione, the Socialist's opposition was successful. His behaviour and the various remarks he made had also cost him the support of the Liberals. Buttiglione resigned his candidature in favour of Franco Frattini to avoid a smouldering row, which would have weighed very heavily on the first days of the Barroso Commission.[197]

In 2009 the EPP again showed its strength by nominating Barroso for a second term. It was no coincidence that an EPP Summit once more played a very crucial role, along with the EPP-ED Group in the EP. Already at the Summit in the spring of 2009, the EPP had backed the candidacy of Barroso: 'The Heads of State and Government of the European People's Party (EPP) today decided to back Barroso for a second term as President of the European Commission. The decision was taken unanimously during the EPP Summit held in Brussels prior to the European Council.'[198] During the electoral Congress on 29–30 April in Warsaw, the EPP officially proposed Barroso as its candidate for the Presidency of the European Commission. It

---

[197] See also Martens, *Europe: I Struggle, I Overcome*, 171–9.
[198] 'EPP Leaders Back Barroso for a Second Term as President of the European Commission' [press release], EPP Summit, Brussels, 19 March 2009.

was the first time a candidacy had been proposed by a European party prior to the European Parliamentary elections. Accordingly, following the EPP's victory in the June elections, the EPP government and party leaders officially proposed Barroso as their candidate on 18 June 2009.[199]

Ministerial Meetings

During the first years of the EPP's existence, most party leaders were happy to rely on those in their parties responsible for European affairs, the international secretaries, for information. Much has changed since then. Conferences of Government and Party Leaders, later on simply called Summits, have created a forum in which leading politicians of the EPP member parties meet with each other regularly. These meetings have been instrumental in EU politics, as well as in the development of the EPP itself. A number of significant successes (for instance the rapprochement with the British Conservatives, institutional reform and Barroso's election as President of the European Commission) have even shaped the course of the EU in general and the EPP in particular.

From 2007 onwards, initiatives have been taken to extend this successful formula to the members of the Council of Ministers affiliated with the EPP. In 2007 the first EPP Foreign Affairs Ministers' meeting took place. These meetings continued in 2008 and 2009 under the leadership of Franco Frattini, Italian Minister of Foreign Affairs, and Elmar Brok, German Member of the European Parliament (MEP). The purpose of these meetings is to discuss, in a private and an informal setting, several issues of foreign affairs. This example was copied by the Ministers of Economy and Finance who started in February 2008 with informal meeting chaired by EPP Vice-President Jyrki Katainen. So far, EPP Ministerial Meetings have taken place in the areas of Foreign Affairs, Economy and Finance (EcoFin), Employment and Social Affairs, Industry, Defence, Agriculture, Environment, Justice and Home Affairs, Transport and Energy. As with the EPP Summits, the EPP acts as a host in these Ministerial Meetings, in order to improve political coordination and policy synergy in the Council. Given the large size of the Council and the importance of EU competences

---

[199] 'EPP Leaders Fully Back Barroso for a Second Term' [press release], EPP Summit, Brussels, 18 June 2009.

for making policies that affect the daily life of every citizen, the EPP wants to take the opportunity to increase the efficiency and effectiveness of EU politics and decision-making, based on the commitment of its leadership and sectorial ministers.

Chapter Nine

The General Secretariat

The General Secretariat is the backbone of the party; it delivers indispensable support for the political leadership of the European People's Party (EPP). The activities of the General Secretariat cover a great variety of tasks: administrative, financial, material and organisational support (particularly for meetings of the Presidency, the Political Assembly, the Summit and the Congress), as well as policy advice in a wide range of areas (competences and countries). It also maintains the EPP's contacts with its member parties and plays an important role in European election campaigns. Based in Brussels, the General Secretariat is composed of a multinational staff and is led by the Secretary General.

Staff

The EPP's General Secretariat began very modestly. For the first few years of its existence, the team consisted of only two people, Josef Müller and Trudi Lücker. It was not until 1981 that two additional members joined the staff. They were Guy Korthoudt, who had previously been Secretary General of the youth organisation of the Flemish Christian Democrats (Christelijke Volkspartij, CVP), and Monique Poket. It soon became clear that the increasing demands made on the Secretariat by member parties were too great for such a small staff. As early as 1979, the electoral campaign for the first direct elections of the European Parliament (EP) had demonstrated the usefulness, for both the leadership and the apparatus of national parties, of having someone to represent their interests at the European level. Thus, the rudimentary structures had to be developed further. The advantages of the Secretariat's existence for the member parties and for the

EPP Group could be demonstrated by the fact that the EPP proved good at organising communications and forming opinion inside the party.

Over the years it became possible, in stages, to double the Secretariat's staff. As a result, the General Secretariat was able to carry out its work, namely the preparation, practical organisation and follow-up of meetings of party bodies in a more satisfactory manner. In particular, the member parties could now receive documents in the languages most important for the EPP's work (German, Italian and French), and it became possible to correspond in a few more languages (Dutch, English and Spanish).

In recruiting new staff, special attention was paid to hiring individuals fluent in several languages and representing a mixture of nationalities. However, for practical reasons, in particular because the office was in Brussels, there was a numerical dominance of Belgians. Over the years the number of staff grew from fewer than 10 to more than 20. As a result of the growing number of member parties, and especially the direct financing from the European Union (EU) budget that began in 2004 (with an important increase since 2007), the EPP was finally able to appoint a significant number of new people. This gave the General Secretariat a new dynamic, particularly since the backgrounds of these new officials varied greatly. In sum, around 25 people currently work in the EPP headquarters.[200] They include the Secretary General, two Deputy Secretary Generals, officials dealing with press and communications, assistants, political advisers and other support staff.

Location

In contrast to the secretariats of the Social Democrat and Liberal European party organisations, the EPP's headquarters have never been based in the offices assigned to the European parliamentary groups in the EP. Instead, the EPP Secretariat was housed in an 'autonomous' location in the centre of Brussels, first on the Rue de la Madeleine, then from 1978 on the Place de l'Albertine, and later (until 1995) on the Rue de la Victoire, where a building was made available by a sympathetic organisation, the Fondation

---

[200] Number of employees as of 1 January 2011, including interns. The number of employees of the Centre for European Studies (CES) is 15, including interns.

Internationale. This building, known as the 'Centro Aristides Calvani', was home to the General Secretariat of the Christian Democrat International (CDI). The arrangement, with separate offices for the EPP and the parliamentary groups, emphasised the party leadership's determination to pursue their tasks in their own way and at their own speed: the intention was to separate party business from the parliamentary routine, and in particular, from the requirements of the EPP Group. Another step was made in 1995 when the General Secretariat moved into its own premises on the Rue d'Arlon, a building closer to the EP and the other EU institutions. Since the number of staff continued to grow and the building became unsuitable for all the activities that had to be organised, in 2006 the EPP left for a totally renovated and much larger building in the Rue de Commerce. The new offices are even closer to the EP building and accommodate all the party's entities, including the member associations. It is now truly the party headquarters and also contributes to the EPP's visibility in EU Brussels. In mid-2010, the Centre for European Studies left the EPP headquarters for a separate building in the same street, Rue de Commerce.

Communication

The General Secretariat has many tasks, but communication has never been the easy one. Not all member parties collaborate as much as they could, and it is very difficult to disseminate the messages of the European political parties through the EU's specialised media, let alone the print and audio-visual media in the Member States. Several attempts have been made, despite a number of thresholds that hamper EU communications in general, such as the language barriers and the absence of a European media market. Some have been successful, others much less so, partly because the EPP for a long time lacked funds and expertise in this area. For instance, in 1983 the *EPP Bulletin: Communiqués from the General Secretariat* was launched to communicate with officials and members of EPP member parties and to inform the general public. It appeared irregularly, documenting or commenting on the key proceedings and decisions of the EPP, such as the results of the political programme being put into practice. In spite of numerous attempts and many good intentions, this bulletin (in six different language editions: German, French, Dutch, English, Italian and Spanish) was not published regularly until 1994. In the early years, the

EPP Group's press department did some extra work for the party. Later, parliamentary work grew and there was increasing public interest in the work of the EP, its deputies and political groups. The extra burden on the press staff in the EPP Group meant that any extra work for the party was limited to special circumstances such as Congresses or Summits of party and government leaders.[201]

Another important publication was the *EPP Yearbook*. Started in 1995, when Klaus Welle became Secretary General, the initiative lasted until 1999. The yearbook gave a very broad synthesis of the EPP's activities, a presentation of the member parties and the representatives in the different party bodies, as well as important party documents such as the full text of the *EPP Newsletter*. Since 2005, the party has started the series again, available in hard copy as well as in electronic form. Nowadays, the Internet makes distribution much easier. All relevant information can easily be found on the party's website: www.epp.eu. In 2008 the party even started its own web TV channel, Dialogue TV. In order to develop the party's communication and media policy in a more sustainable and professional way, a spokesperson has been appointed in addition to the party's press officer. The spokesperson, elected by the Political Assembly for a period of three years, is responsible for 'promoting the profile and the work of the EPP and the members of the EPP Presidency in all the media and related public fora'.[202]

Secretary General

The Secretary General implements decisions reached by the party bodies. With the support of the General Secretariat, the Secretary General prepares meetings and is concerned with cooperation and agreement with and among member parties, the associations and the parliamentary groups.[203]

---

[201] In terms of numbers of officials, there is a huge gap between the party and the EPP Group, especially with regard to communication staff. The EPP Group has a staff of press counsellors who cover all Member States and/or language groups. Together they total around 35 people (figures as of 1 January 2011).

[202] EPP Internal Regulations approved by the EPP Congress on 10 December 2009 in Bonn, title IX.

[203] EPP Statutes approved by the EPP Congress on 10 December 2009 in Bonn, article 19.

For a long time, the Secretary General was elected by the Political Assembly (previously the Political Bureau), on the President's recommendation. Since 1999 the Secretary General has been elected, as has the President, by the Congress. The same person could also be Secretary General of the European Union of Christian Democrats (EUCD), as was the case from 1983 until the merger between the EPP and the EUCD to coordinate the common action of Europe's Christian Democrats beyond the boundaries of Community countries.

The first Secretary General of the EPP, Jean Seitlinger, held an honorary post. During Seitlinger's entire term of office, from 1978 to 1983, he was both a Deputy in the French National Assembly and a Member of the European Parliament (MEP); he was also Mayor of Sarreguemines in Lotharingen. He left the daily managing of the General Secretariat to the Executive Secretary (and later acting Secretary General), Josef Müller. A former Member of the German Bundestag and of the European Parliament, Müller had broad political experience, particularly in European matters.[204]

This situation was generally held to be unsatisfactory. The appointment of a new Secretary General was seen as a first step towards breathing life into the EPP structure as it had been envisioned at the time the party was founded. The holder of this position should not be burdened with parliamentary duties, but be available to devote full attention to the job. Helmut Kohl, the federal leader of the German Christian Democrats (Christlich Demokratische Union, CDU), insisted to EPP President Leo Tindemans – who, according to the statutes, had the right of proposal – that the position must be filled by a German. The new Secretary General of the EPP should at the same time take over responsibility for the EUCD General Secretariat, which was to be transferred from Rome to Brussels and integrated with the EPP General Secretariat.

Thomas Jansen was elected in April 1983 and remained Secretary General of both the EPP and the EUCD until the end of 1994. He served under all EPP Presidents: Tindemans, Piet Bukman, Jacques Santer and Wilfried Martens. Before he took up this position in Brussels, he led the office of the Konrad Adenauer Foundation in Rome, acquiring among other things a

---

[204] Since losing the Aachen constituency Bundestag seat in 1972, Müller had, on behalf of the CDU, busied himself with establishing an office in Brussels whose job – working closely with the EPP Group in the EP – was to coordinate the work of the EUCD parties at Community level. This office was simultaneously the EUCD's Brussels office, the Secretary General at that time still being based in Rome.

great knowledge of and sensitivity to Italian politics, unlike many other EPP politicians. After serving as Secretary General, Jansen worked in the Forward Studies Unit of the European Commission, the study centre of Commission Presidents Santer and Romano Prodi. Before his retirement in 2004 he served as head of the President's Office of the European Economic and Social Committee (EESC).

In Jansen's opinion, the Secretary General had to be free to defend the interests of the EPP against the demands of individual member parties and the EPP parliamentary Group, and against the ambitions of individual party members. The mandate he was given meant that he was to serve solely the EPP, devoting himself entirely to further developing the party and to organising its work. By contrast, all other members of the Presidency, including the President, normally have additional duties arising from their other positions or mandates. These other duties – such as obligations to their national parties, to their associations or to their parliamentary group – will usually take precedence. The Secretary General should embody the party's conscience and therefore serve only the EPP. In any event, the individual holding this position should be the conceptual and dynamic force behind all undertakings intended to give the party a public profile, whether organisational or in terms of its political programme.

Between 1994 and 1999, Welle was Secretary General of the EPP and the EUCD. As Head of Foreign and European Affairs for the CDU in Bonn (1991–94), he was closely involved in EPP activities, especially in drafting the 1994 election manifesto. His term as Secretary General saw the EPP enlarge to include numerous Conservative parties from old and new Member States, an operation Welle wholeheartedly supported and facilitated. It was also during his term that the Youth of the European People's Party (YEPP), the European Senior Citizens' Union (ESCU) and the Small and Medium Entrepreneurs' Union (SME UNION) were founded. As Secretary General he had relatively more leeway than his predecessor, since President Martens was EPP Group Chair between 1994 and 1999. In 1999 Welle left the EPP to become Secretary General of the EPP-ED Group, a position he held until 2004. He then became a civil servant in the EP's General Secretariat, successively as Director General for Internal Policies, Head of Cabinet during Hans-Gert Pöttering's tenure as President of the European Parliament, and Secretary General.

From 1999 until 2002 Alejandro Agag Longo was EPP Secretary General. Previously, he was involved in the youth organisation of the Spanish

People's Party (Partido Popular, PP), and was a personal adviser to Prime Minister José Maria Aznar. Agag was succeeded by another designated aide to Aznar, António López-Istúriz. López was re-elected as Secretary General in 2004, 2006 and 2009. Both appointments emphasised the strong position the PP had acquired within the EPP. Unlike their predecessors, both Agag and Lopez became MEPs. This created stronger ties with the Group in the EP, but meant that they could not be as present as their predecessors. This reinforced the position of the EPP President and, inside the General Secretariat, the position of the Deputy Secretary General.

Deputy Secretary General

On the proposal of the Presidency and in agreement with the Secretary General, the Political Assembly can elect a number of Deputy Secretary Generals for a term in office of three years. The EPP veteran Korthoudt was the first Deputy Secretary General. After having worked in the General Secretariat, he became Deputy Secretary General in 1986. He took on the organisational, financial and administrative work. In this position he was succeeded in 2004 by another Belgian from the Flemish Christian Democrats (Christen-Democratisch & Vlaams, CD&V), Luc Vandeputte. Following Welle's departure as Secretary General in 1999, a second Deputy Secretary General, Christian Kremer from the CDU, was appointed to take on political duties under the authority of the party president. Kremer has been, among other things, responsible for drafting EPP manifestos, for the representation of the EPP in the European Convention and for a number of EPP working groups. After the establishment of clear responsibilities in the General Secretariat, Kremer became Head of the Policy and Strategy Department while Vandeputte became Head of the Organisation, Finances and Human Resources Department. To adequately deal with the 'EPP's bilateral and multilateral relations with like-minded parties and organisations beyond the EU and in other continents', a Secretary of External Relations has been appointed.[205] This position has been filled by the Greek Kostas

---

[205] EPP Internal Regulations approved by the EPP Congress on 10 December 2009 in Bonn, title VIII. The Secretary of External Relations is elected by the Political Assembly for a period of three years.

Sasmatzoglou from 2005 to 2009, and since 2009 by Nicolas Briec from France.

Treasurer

Unlike the Secretary General and his Deputies, the function of Treasurer has never been part of the EPP staff. The Treasurer is part of the Presidency and is elected according to the same rules that apply to the election of the Vice-Presidents. The responsibility of the EPP Treasurer is to work out a budget with the Secretary General and the Deputy Secretary General responsible for financial affairs and present it to the Political Assembly, to make sure member parties pay their dues, to ensure budgetary decisions are executed and to guarantee that the party is eligible for European party resources. During the early years of the party, the position of Treasurer was held by two enormously reliable people, both Flemish Belgians: Alfred Bertrand (1978–85) and Rika De Bakker (1985–96). It proved sensible to give this position to Belgian politicians, as this simplified dealings with the banks and authorities of Belgium, where the General Secretariat was based. In 1996 Ingo Friedrich became EPP Treasurer. Like his predecessors he was until 2009 an MEP and therefore his position strengthened the ties between the party and the Group in the European Parliament. Friedrich is a Deputy Chair of the Bavarian Christian Social Union (Christlich-Soziale Union, CSU) and one of the few politicians to have been a member of the EP since the first direct elections in 1979.

Chapter Ten

Working Groups and Member Associations

Working Groups

Although they are not an official party body, Working Groups have been an integral part of the European People's Party (EPP) for a long time. They are the backbone of its political work, where high-profile representatives from EPP member parties develop common positions and strategies on major policy areas, and submit specific recommendations to the Political Assembly for final approval. A Working Group may be given a specific mandate for preparative work for an EPP Summit or Congress. A Working Group is normally chaired by the president or one of the vice-presidents. The appointment of chairs is the responsibility of the EPP Presidency.[206]

In the period 2010–2012, permanent Working Groups exist for the following areas: European Policy – including institutional reform of the European Union (EU) – (WG1), co-chaired by Wilfried Martens and Peter Hintze; Economic and Social Policy (WG2), co-chaired by Gunnar Hökmark and Rumiana Jeleva; and EPP Membership (WG3), chaired by Corien Wortmann-Kool.[207] A Working Group on Foreign and Security Policy existed until the 2006 Rome Congress, but its mandate is now dealt with by the Working Group on European Policy. Traditionally this Working Group drafts the EPP electoral manifesto. Other Working Groups may also be in-

---

[206] EPP Internal Regulations approved by the EPP Congress on 10 December 2009 in Bonn, title VI.
[207] Gunnar Hökmark and Rumiana Jeleva succeeded Ria Oomen-Ruijten and Peter Jungen; Wortman-Kool succeeded two other Dutch MEPs and EPP Vice-Presidents: Wim van Velzen and Camiel Eurlings.

volved in drafting the EPP programme. The Working Group on Economic and Social Policy, for instance, drafted the 2009 Bonn Congress document on the social market economy and the resolution on the reform of the financial markets. With the EU dealing increasingly with social, economic and financial affairs, the importance of this Working Group has grown.

Although the EPP is now represented in almost all European countries, the Working Group EPP Membership, previously called 'Central and Eastern Europe' and 'Enlargement', continues to receive applications from different countries within and beyond the EU. In recent years cooperation between the EPP and parties from countries of the Eastern Partnership (Moldova, Georgia, Armenia, Azerbaijan, Ukraine, Belarus), as well as with those of the Balkan region, has improved. In response to applications from these countries, the Working Group organised fact-finding missions (Moldova and the Czech Republic) and study trips (Serbia, Bosnia and Herzegovina). It also launched workshops as a vehicle to exchange views and share knowledge between experts and Working Group members in different areas of external relations.

Next to the three permanent Working Groups, three ad hoc Working Groups have been established: one on Agriculture (WG4), co-chaired by Wilhelm Molterer and Piet Bukman; one on Climate Change (WG5), chaired by Karl-Heinz Florenz; and one on Eastern Partnership and the Mediterranean Union (WG6), co-chaired by Thierry Mariani and Mario David. As well as the Working Groups, there is the Campaign Management Group, chaired by Deputy Secretary General Christian Kremer, which normally meets twice a year, and more often during election periods. The purpose of the meetings is to discuss common strategies, for instance, on preparing for European elections and, more importantly, to provide a platform where member parties can discuss their recent campaigns and exchange information on campaign techniques.[208]

Member Associations

Unlike many of the other European parties, the EPP has a long history with Member Associations, and they have been relatively closely involved in

---

[208] Interview with Deputy Secretary General Christian Kremer, Brussels, 17 December 2009.

the party's activities.[209] No fewer than seven Associations are affiliated with the EPP: the Youth of the European People's Party (YEPP), the EPP Women, the European Union for Christian Democratic Workers (EUCDW), the Small and Medium Entrepreneurs' Union (SME UNION), the European Senior Citizens' Union (ESCU), the European Democrat Students (EDS) and the European Association of Locally and Regionally Elected Representatives (EALRER). They are officially recognised as Member Associations by the Political Assembly. One of the requirements for recognition is that an Association must be composed of national sections that are linked to an EPP member party in at least half of the EU Member States.[210] Recognition gives Member Associations the right to participate in party bodies, as well as access to substantial grants from the party budget. Yet provided they operate in accordance with the statutes and the party programme, the associations remain largely autonomous as far as their internal affairs and activities are concerned.

Apart from the women's and youth organisations, not every EPP member party has associations that bring certain categories of members together to work for the party in their fields, or alternatively to press their particular causes. This model of internal party associations had no real historical basis or deep roots outside of the German, Austrian and Belgian member parties. The EUCDW, the EALRER and later the ESCU came into being because of initiatives and pressure from Belgian and German organisations. Moreover, with the exception of the SME UNION, all EPP associations were also European Union of Christian Democrats (EUCD) associations. Originally the women's and youth groups had separate organisations for their activities in the EUCD and EPP. In time, the existence of parallel organisations proved unnecessary, and as the EU gradually expanded, it proved downright superfluous; enlargement tended to make the EPP's and the EUCD's spheres of operations one and the same.

The EPP's interest in the founding and the activities of such associations resulted from the fact that they substantially increase the number of people and groups directly involved in European work. In turn, people engaged in

---

[209] Since the 2009 Bonn Congress they are no longer called 'Recognised Associations' but 'Member Associations'. See EPP Internal Regulations approved by the EPP Congress on 10 December 2009 in Bonn, title X.

[210] Ibid.

one of these associations improve their knowledge and experience about European politics. Connections are forged and opportunities created for building trust and solidarity among EPP members from different countries, which has a direct effect on the member parties.[211] In this way, a core membership of the 'European party' can be permanently established. Equally, the involvement of people representing a particular category of member or a particular interest group enlivens and enriches party work. These people want to spread understanding and to exercise influence – to make a difference in the parties they belong to. The contributions of the associations, generally concentrating on the specific interests of their supporters, provide a link between the real world and the consensus towards which the party is always striving. In influencing EPP proposals, this role is of great importance for a people's party whose power is derived from being able to unite divergent interests.

Youth of the European People's Party (YEPP)

As early as the 1950s, there was a European or international youth organisation. It was originally connected to the Nouvelles équipes internationales (NEI), then to the EUCD and to the Christian Democratic World Union (CDWU). Once the EPP was founded, the youth organisation adapted so that it could be active in this new area.[212] But until 1997, when the YEPP was formed, three organisations existed alongside each other.

When the EPP was founded, a 'Team of the Nine' was set up as an EPP association under the framework of the European Union of Young Christian Democrats (EUYCD). When Greece joined the European Community (EC) it became the 'Team of the Ten'. The EPP association was called the European Young Christian Democrats (EYCD). The EUYDC, not to be confused with the EYCD, was the EUCD's association. In the early 1980s, when the EYCD was reformed, the two structures were integrated.

Every youth organisation is subject to changes and discontinuity in terms of personnel, and this was certainly one of the reasons for the

---

[211] In turn, Member Associations receive financial support if they meet a number of requirements. See EPP Internal Regulations approved by the EPP Congress on 10 December 2009 in Bonn, title XVI, part g.

[212] One important indication of its influence was that in 1994, for the first time, an EYCD representative, Secretary General Marc Bertrand, was elected an EPP Vice-President.

EYCD's phases of minimal or inadequate activity. But the EPP's young people were always dynamic, and made a considerable contribution to spreading an EPP consciousness to the youth organisations of member parties. A large number of young people who had their first European experiences in the EYCD or the EPP remained active in European politics when they were older, some of them becoming involved in the EPP's structures or in areas close to the party. The most prominent example so far is Swedish Prime Minister and Chair of the Moderate Coalition Party (Moderata Samlingspartiet or Moderaterna, MS), Fredrik Reinfeldt, who chaired the Democratic Youth Council (DEMYC) from 1995 to 1997 and was the first YEPP President (1997–99).[213]

Given the difficult conditions and limited resources of the EYCD, it carried out consistently good work over the years. This is especially remarkable given that there were fierce, constantly recurring battles inside the organisation, ignited by problems which also led to controversies within the EPP. There were three main areas of argument: the double membership issue, the inclusion of Conservative party associations and the question of who had the right to be represented in party bodies. These questions were either objectively or formally interlinked, which meant that the front lines in these battles were almost always intertwined too.

The issue of double or overlapping membership resulted from the fact that some member associations, particularly the youth organisation of the German Christian Democrats (Christlich Demokratische Union, CDU), Young Union (Junge Union, JU), also belonged to the DEMYC, an organisation that contained both Christian Democratic-inclined and Conservative groups. The dual membership corresponded with the CDU and the Bavarian Christan Social Union (Christlich-Soziale Union, CSU) belonging to both the EPP and the European Democratic Union (EDU). For those opposed to joining forces with the Conservatives, the situation was clear: every political youth organisation had to choose between the EYCD and DEMYC. If it was no longer possible to force the JU and the youth organisation of the Austrian People's Party (Österreichische Volkspartei, ÖVP) and other founding members to renounce their DEMYC membership, at

---

[213] Reinfeldt succeeded Klaus Welle, who chaired the DEMYC from 1991 to 1994. Earlier on, Elmar Brok, also from the Young Union (Junge Union, JU) and now one of the leading MEPs of the EPP Group, was Chair between 1979 and 1981.

least it should be made a condition that any new member could join only if they gave up the right to dual membership.

Quarrels had existed among the EYCD about bringing in the Conservatives for as long as anyone could remember. The founding of the DEMYC itself was in a certain sense the result of a compromise. In the 1970s it had become clear that there was a stalemate. Several powerful groups (e.g. the Italian Youth Movement (Movimento Giovanile), the youth organisation of the Belgian Christian People's Party (Christelijke Volkspartij, CVP) and of the Dutch Christian Democratic Appeal (Christen Democratisch Appèl, CDA) were against bringing the Conservatives on board, arguing that it would be impossible to integrate them into a European federation of young Christian Democrats. It was equally impossible to dissuade proponents of integration (notably the Germans and the Austrians) to abandon closer co-operation with the Conservatives.

Another problem arose because of an under-representation of the JU, which the association felt to be unfair. It was by far the largest and politically most powerful member association but felt constantly outnumbered by a series of smaller associations taking radical positions on the problems mentioned above. Insofar as the EPP supplied, from the beginning of the 1990s onwards, a solution to the 'Conservative question', the EYCD could gradually get back on an even keel. This could potentially lead to 'the creation of a new EPP youth organisation, open to all EPP/EUCD member party youth organisations' and to the

> acceptance of a statute for the new EPP youth organisation stating in which it would be stipulated that members were not allowed dual membership of any other youth organisation (specifically not the DEMYC). The voting strength of member organisations would be laid down (this being based on the political importance of the mother party, and the country's population, as well as the organisation's activity, along with recognition of the basic programme and the [EPP] statutes.[214]

A gradual rapprochement took place between the EYCD and DEMYC which eventually led to the founding of the YEPP during the Congress in Brussels on 31 January 1997. Since then, the EPP's political youth organisation has grown steadily in terms of membership and activities. It now brings together more than 50 political youth organisations, predominantly

---

[214] 'The Future of the European Young Christian Democrats: The Youth Organisation Faces Change of Course' [interview with Marc Bertrand], *EPP News* 24 (1995).

from the centre-right, from 38 countries all over Europe.[215] Its mission is 'to maintain good contacts within our political family [the EPP], provide ground for training, discussion and cooperation and prepare the younger generation to lead tomorrow's Europe. This runs parallel to our mission to stimulate further integration in Europe, to uphold our principles, policies and ideas in the European political debate.'[216] Since Reinfeldt's term, six people have chaired the YEPP: Michael Hahn (1999–2001), Rutger-Jan Hebben (2001–03), Daniel Bautista (2003–05), David Hansen (2005–07), Ioannis Smyrlis (2007–09) and Laurent Schouteten (since 2009).

The EPP Women

The EPP Women have also developed a fairly good working relationship with the party, but unlike the YEPP, they are less at the forefront of EPP affairs and less organised. As early as 1981, five years after the party's founding, a women's section was established. This organisation existed alongside the one from the EUCD for a long time. In 1992 they were merged at a Congress of both organisations, reflecting what had largely already taken place in practice. The necessity of uniting the two associations was clearly underlined by the impending expansion of the then EC, a definite prospect from 1989 onwards. The merger meant that the apprehensions of the women whose countries were not (or not yet) members became less important. Rivalry between the associations, evidently mixed up with rivalries between certain leading figures, also fell away to a great extent.

The last President of the EUCD's women's organisation was Concepcio Ferrer. Its leading figures have always been active in EPP party bodies. The Chair, Marlene Lenz, was elected an EPP Vice-President in 1994. From 1996 to 2000 the Chair was Agnès van Ardenne-van de Hoeven from the Netherlands. The current Chair is German Member of the European Parliament (MEP) Doris Pack. Generally speaking, the EPP Women are 'dedicated to the advancement of women's political emancipation throughout Europe and the promotion of important women-related issues'. It counts about 40 political women's organisations from 38 countries in its membership.[217]

---

215 Figures as of 1 January 2011. See www.yepp-eu.org.
216 Ibid.
217 Figures as of 1 January 2011. See www.epp-women.org.

European Union of Christian Democratic Workers (EUCDW)

Given its large national memberships, the association of Christian Democratic workers ought to have significant potential as an EPP association. In practice, it has been less successful than the youth organisation. Founded in the turbulent year of 1980, and despite being led by such influential figures as Alfred Bertrand and Hans Katzer, the EUCDW found it too hard, for too long, to show the kinds of results that were expected of it. The organisation suffered above all from leadership problems that arose from unfortunate coalitions; these in turn could be traced to a lack of consensus about the association's purpose.

The problems faced by the youth organisation, notably the 'Conservative question', did not afflict the EUCDW, which was always solid on the issue. Moreover, there were no worker organisations in the Conservative camp seeking to join the EUCDW. The problems of Christian Democratic-inclined trade union activists belonging to EPP parties were mostly found on other levels of decision-making. Their national associations and their members saw a permanent conflict of solidarity between trade union and party. Indeed, it was the flip side of the conflict faced by the EPP employers' federation. Ideally, the organisation should represent the party in the unions, and the unions in the party. But what happens in practice in each country depends on national structures, and is very different in each case. The alliances in the EUCDW arose, after all, from the different identities and priorities of national associations, and their closeness or otherwise to their union organisations and their parties. These difficulties were overcome after 1992, thanks to the efforts of Heribert Scharrenbroich, Jean-Claude Juncker and Miet Smet.[218] In 1997 Smet was succeeded as EUCDW President by the Belgian Luc Delanghe. On 5 October 2002 at the EUCDW Congress in Malta, German MEP Elmar Brok was elected as the new President.

The EUCDW's way of working largely corresponds to that of the YEPP. The leading bodies meet regularly to deal with practical cooperation and organisation, and to discuss and decide on current political issues and po-

---

[218] Scharrenbroich, President of the International Union of Christian Democratic Workers, took over as leader after a critical development in 1992; in 1993 Juncker took over the Presidency until his election as Prime Minister of Luxembourg; his successor from early 1995 was the then Belgian Minister for Labour and Employment Miet Smet.

licy directions; the EUCDW also holds seminars. There are congresses every two years to hammer out political directions. The issues that especially interest Christian Democratic employees are in the field of community and social policy. However, they are traditionally interested in politics as a whole and are especially committed in terms of European policy. The organisation and financing of educational work – which mainly consists of running seminars for members and others involved with companies and trade unions – is done through the European Centre for Training Employees, set up for this purpose by the EUCDW. The EUCDW currently brings 24 member organisations from 18 countries together.[219]

The Small and Medium Entrepreneurs' Union (SME UNION)

The European Small and Medium-sized Businesses Association, as it was known when it was founded in 1980, always confined itself to being a forum for discussion. Executive President Lieven Lenaerts was for many years the main force behind and organiser of the association. His main concern was to ensure that the views and interests of politicians representing small- to medium-sized enterprises were included in EPP discussions. At the same time, he sought to offer a service to the association's members by holding educational seminars and by supplying contacts and information. One of the Association's problems, however, was too few member parties in the EPP with comparable organisations of their own, the result of which was a lack of branch organisations and, therefore, lack of an active base.

When the Association appeared to be at a standstill, it was hoped that a planned merger with the European Association of Medium-sized Enterprises would overcome the problem.[220] The latter group was founded in 1979 to represent the political interests of small- and medium-sized entrepreneurs, the self-employed, artisans and skilled employees.[221] Its political direction was always Christian Democratic and Conservative. However, it wanted to avoid, at least during its first year of existence and certainly officially, being a party association. The European Association of Medium-sized Enterprises recruited members from non-party or principally

---

[219] Figures as of 1 January 2011. See www.eucdw.org.

[220] See 'EMSU and EMV Plan Fusion', *EPP News* 13 (1995).

[221] One of the founders was Wolfgang Schüssel, later to become ÖVP Chair and Chancellor of Austria.

apolitical professional associations of small- to medium-sized businesses. Despite this, and not in the least because of the figures who headed the organisation, there were always close ties with the EPP and in particular with the EDU, which for some years regarded the European Association of Medium-sized Enterprises as a de facto EDU association.

In 1996 both organisations merged into the EPP Independent Business and Economic Foundation, later known as the SME UNION. The SME UNION's 'top priority is to reform the legal framework for SMEs all over Europe and to promote and support the interests of small and medium-sized enterprises'.[222] After many years as Chair of the EPP's SME network, Lenaerts was succeeded by Marianne Thyssen (1996–98), Karla Peijs (1999–2000), Paul Rübig (2001–03), Jacques Santer (2003–05) and Christoph Leitl (2005–09). Since 2009, Jungen has been the Chair of the SME UNION.

The European Senior Citizens' Union (ESCU)

From 1992 onwards, efforts were made to form an EPP association dedicated to pensioners. The rediscovery of the 'third age' has meant that some member parties offer older people the opportunity to do party work within special associations. The success of these initiatives and the increasing need for a new policy to help older people eventually led to these national associations being brought together at the European level. The foundations of a European political organisation were laid during a pensioner's conference in May 1993 that was organised by the EPP Group in the European Parliament (EP) with the cooperation of the Konrad Adenauer Foundation.

After many false starts, a workshop was organised in Aachen on 5 May 1995. It had the purpose of preparing for the founding of an association within the framework of a Congress.[223] The very small number of people in this workshop, coming from Belgium, Germany, Finland, Luxembourg, Austria and Spain, indicated a particular difficulty with pensioners. If there was a lack of national structures embracing small- to medium-sized businesses, local politicians or employees, the obstacle was far more pronounced in the case of the pensioners. Nonetheless, efforts to establish an association continued to be made. They eventually led to the founding of the ESCU in Madrid on 7 November 1995, in the course of the EPP Con-

---

[222] See www.sme-union.org.
[223] See 'European Pensioners Union Workshop Founded', *EPP News* 14 (1995).

gress. The programme of the ESCU was laid down in the Declaration of Vienna.[224] Since then, a number of congresses and other activities have been organised. Similar to other associations, the ESCU has been active in drafting and publishing reports, statements and other documents, at EPP Congresses and other places. From its founding until 2001, the ESCU was chaired by Stefan Knafl. Since 2001, Bernhard Worms has been the Chair.

The European Democrat Students (EDS)

The EPP's student organisation was founded in 1961 in Vienna as the International Union of Christian Democratic and Conservative Students. In 1970 the name was changed to European Union of Christian Democratic and Conservative Students. In 1975 the name was changed again, to EDS, to reflect all tendencies within the organisation, and to focus on the need for cooperation among those European forces that believe in an open, democratic and non-Socialist society.[225] It was the organisation's Chair, Carl Bildt, later to become Chair of MS and Swedish Prime Minister, who proposed a more neutral name to emphasise the organisation's objective to establish a (centre-right) 'European Democrat Party'.[226]

The organisation has always been very active in promoting European integration, defending students' interests and looking for partners within and beyond the students' community and the area of higher education. This is unusual, since it has predominantly been led by British and Nordic students. Gradually becoming the main representative of centre-right–oriented student organisations all over Europe, including Central and Eastern Europe, the EDS was eventually recognised as an EPP Association by a decision of the Political Bureau, now the Political Assembly, on 9–11 November 1997. Since then, the EDS has not stopped expanding in terms of member organisations and in terms of its activities, such as lobbying, training and sharing information.

Since 1997 no fewer than 10 students have chaired the EDS: Günther Fehlinger (1996–98), Michalis Peglis (1998–99), Ukko Metsola (1999–2000), Gustav Casparsson (2000–01), Jacob Lund Nielsen (2001–03),

---

[224] Walter Paul, 'History of the Origins of the ESCU', http://www.eu-seniorunion .info/en/background/ESU_origin-en.pdf.

[225] EPP, *Yearbook 1999* (Brussels, 2000), 343.

[226] See http://www.edsnet.org/about-us/history for a more extensive history of the EDS.

Alexandros Sinka (2003–05), Sven Henrik Häseker (2005–06), Ana Filipa Janine (2006–08), Thomas Uhlen (2008–09) and Bence Bauer (since 2009). Currently, EDS has 37 member organisations representing nearly 500,000 students in 32 countries, including non-EU member states like Belarus and Georgia.[227]

The European Association of Locally and Regionally Elected Representatives (EALRER)

Local politicians have always been among the staunchest defenders of European integration. The network of twinned cities and community partnerships has done a great deal to breathe life into the European idea, as well as to include Europe's citizens in the integration process. The Council of European Regions, part of the International European Movement, has coordinated these efforts and also given them a political conduit. So it was natural enough that Christian Democratic-inclined local politicians working together in the Council of European Regions felt the need to institutionalise their relationships through an EPP association. This came about at the beginning of the 1980s with the founding of the European Federation of Local Politicians (EFLP). No doubt it was hoped or assumed that establishing an EPP group in the Council of European Regions would strengthen the EPP's position vis-à-vis the Socialists, who were especially strong in this institution.

The EFLP tried, initially with some success, to develop a network of local politicians from various member parties who were involved in European politics. However, in practice, the mainstream European issues were being debated and decided in the usual EPP bodies. Finding a clear area of activity at European level that local politicians could make their own was not easy. Given the subsidiarity principle, held especially dear by local politicians, formulating a regional policy for Europe was hardly in their interests. And European policy from a local political perspective could be nothing other than a general European policy, which is not exclusively up to local or city councils to formulate.

So the work of the EFLP could not really develop beyond useful contacts and agreements. It became the local political conscience of the EPP. In that respect, the EFLP was thoroughly effective. It channelled the local political point of view into the process of elaborating party programmes,

---

[227] Figures as of 1 January 2011. See www.edsnet.org.

and it also spread the EPP's message at the local political level. But that was evidently not enough of a basis for an association of active politicians with real work to do at home. The absence of any definite project relevant to their actual problems necessarily meant that local politicians never became genuinely interested in the association. Partly because of this, there were problems with the leadership. Both President Leoluca Orlando (elected in 1985), then Mayor of Palermo, and Secretary General Peter Daners completely neglected the organisation. The member associations were inert, in effect leading to the de facto extinction of the EFLP.

The association came back to life after 1991, thanks to the further development of the EU, as expressed in the Maastricht Treaty. Subsidiarity was recognised as a key principle of the EU, a fact that for the first time gave constitutional recognition to the role of regional and local politics in shaping the Union. This had the effect of mobilising people and made it possible, after detailed preparations during the spring of 1993, to set up the successor organisation to the EFLP. This new organisation was called the European Local and Regional Political Union (ELRPU). Adolf Herkenrath, who had already been a driving force in the CDU/CSU's association of locally elected people (Kommunalpolitische Vereinigung), and who had taken the initiative of reforming the association, was elected President of the ELPRU. François Biltgen, who had done most of the practical work in the run-up to refounding the association, became Secretary General.[228]

This new organisation was more promising than its predecessor, because the Maastricht Treaty raised the profile of communes and regions as necessary elements of a European framework. The Committee of the Regions (CoR) set up in March 1994 as a new European institution to involve the communes and regions in forming political will, also became an important arena for the ELRPU. However, the organisation has not always lived up to its potential, despite changing its name to European Association of Locally and Regionally Elected Representatives (EALRER) and is therefore currently waiting for a new start.

---

[228] Since 2003 Biltgen has been Chair of the Luxembourgisch Christian-Social People's Party (Chrëschtlech-Sozial Volekspartei, CSV) and has served as a minister in governments led by Jacques Santer and Jean-Claude Juncker.

Chapter Eleven

Statute and Financing

Political parties are essential to democracy. Democracy is hardly imaginable without the notion of party democracy. Nevertheless, at the European level, political parties are relatively new, compared for instance with the party groups in the European Parliament (EP) and its forerunner, the General Assembly. It took a while for European political parties to be officially founded; it was another political generation before their role was eventually enshrined in the Treaties. As a result of the Treaties, since 2004 European parties have at their disposal some essential resources to develop activities inside and outside Brussels. The work is not finished, however, as the road towards a truly European party system is bumpy and long.

The Founding of European Parties

At an early stage, the main political parties in the six founding states of the European Union (EU) had begun to cooperate with like-minded parties in other countries of the European Community (EC). From the late 1940s through the 1950s, European party families began to form. They reached agreements among themselves and increasingly also acted in concert. One fruit of this gradual integration came in the mid-1970s, with the prospect of the first direct elections to the EP in 1979. This prompted the establishment of the first properly constituted federations of parties. The challenge of the European elections pushed the traditional political families to set up pan-European organisations. Notable enthusiasts were the Members of the European Parliament (MEPs) who had formed parliamentary groups since the early 1950s, as they felt an increasing need to be able to rely on 'European parties'.[229]

In this way, in 1974 the Confederation of Socialist Parties in the European Community was created from the Liaison Bureau (established by the Socialist International in 1957) to further cooperation between EC parties. The Liberal International decided in 1972 to examine ways of deepening cooperation between parties in the EC. In 1976 the Federation of Liberal and Democratic Parties in the European Community was formally founded; but in contrast to its Socialist counterpart, it was independent of the Liberal International. In the same year, the European People's Party–Federation of Christian Democratic Parties in the EC was founded.[230] Not surprisingly, these organisations were modelled after their national member parties: a congress of delegates decides on the political programme, an executive committee deals with current issues and day-to-day business, a chair (supported by a party presidium or board) speaks for the party and represents it, and a secretary general (supported by a secretariat) is in charge of internal communications and the technical and organisational work necessary to ensure the party bodies can operate properly.[231]

With the prospect of direct elections, the national parties had – for the first time – a clear interest in running a coordinated European election campaign. The national parties also hoped to benefit from the publicity that might possibly accrue from belonging to a supranational organisation. The opportunities afforded by this kind of confederation in Community politics were obvious, even at this early stage. But the very meaning of the establishment of European parties went beyond the EP elections. The transnational cooperation institutionalised in the European parties also had its effect on the mentality and behaviour of the leading figures in national

---

[229] See Wolfgang Wessels, *Zusammenarbeit der Parteien in Westeuropa. Auf dem Wege zu einer neuen politischen Infrastruktur?* Institute for European Politics (Bonn, 1976), and Theo Stammen, *Parteien in Europa. Nationale Parteiensysteme. Transnationale Parteienbeziehungen. Konturen eines europäischen Parteiensystems* (Munich, 1977).

[230] Steven Van Hecke, 'On the Road Towards Transnational Parties in Europe: Why and How the European People's Party Was Founded', *European View* 3 (Spring 2006), 156. See also chapter two of the present volume.

[231] For information on the activities and development of European parties, see the annual contributions on party federations published since 1980 in Werner Weidenfeld and Wolfgang Wessels (eds.), *Yearbook of European Integration* (Bonn, 1980–).

party politics. To political parties operating at Union level, it became obvious that parties needed to be present at Union level if they were to look after their interests there, wield an influence or actively help shape the political architecture of Europe.[232] The harmonisation process between the different member parties that was necessary to form larger federations also had implications for the identities of national parties – on how national parties and/or their leaders saw themselves and how they were to sell their common ideas beyond party structures. In turn, intensive work done on political programmes strengthened the ties between the national parties, the MEPs and the political families to which they all belonged.[233]

In 1977, for instance, with all their MEPs sitting together in the Christian Democratic Group in the EP, the three traditional Christian parties of the Netherlands found that the harmonisation process was encouraged by their cooperation at EP level, and facilitated by the imminent creation of a European party of Christian Democrats. The Catholic People's Party (Katholieke Volkspartij, KVP), the Christian Historical Union (Christelijk-Historische Unie, CHU) and the Anti-Revolutionay Party (Anti-Revolutionaire Partij, ARP) felt impelled to join forces. A year earlier, all three had been co-founders of the European People's Party (EPP). Their joining together produced the Christian Democratic Appeal (Christen Democratisch Appèl, CDA).

An Initiative by European Party Leaders

More or less systematic cooperation between like-minded parties brought with it progressively more elaborate organisational and communication structures. If European parties were to become more than just a vehicle to prepare for EP elections, they should be able to access the necessary resources (institutional and material) to function as parties in a political system

---

[232] Oscar Niedermayer, *Europäische Parteien? Zur grenzüberschreitenden Interaktion politischer Parteien im Rahmen der Europäischen Gemeinschaft* (Frankfurt am Main/New York, 1983).

[233] See Martin Bangemann et al., *Programme für Europa. Die Programme der Europäischen Parteienenbünde zur Europawahl 1979* (Bonn, 1978); Eva-Rose Karnofski, *Parteienbünde vor der Europawahl 1979* (Bonn, 1982). Cf. Eberhard Grabitz and Thomas Läufer, *Das Europäische Parlament* (Bonn, 1980), 295ff.

such as the then EC. It was logical, in this context, that the representatives of the European parties took the lead. After all, they had a clear interest in the development of a truly European and functional party system.

In 1989, the Secretary Generals of the three party federations had begun meeting from time to time to talk about common problems and to exchange experiences. The result of these conversations was the idea of bringing their party leaders together and starting joint discussions on 'the development and role of European parties or party federations in the Community's political system', and on 'relations with the groups in the European Parliament'.[234] The first meeting of Wilfried Martens, Guy Spitaels and Willy De Clercq – the Chairs of the EPP, the Confederation of Socialist Parties of the European Community and the Federation of European Liberals and Democrats respectively – took place on 18 September 1990. They agreed to talk again and to hold a joint press conference just before the European Council met in Rome on 12 December 1990 to convene the Intergovernmental Conference (IGC) on the Treaty on European Union. The three explicitly wanted to make a joint statement to ensure their demand for a role for the European parties was firmly on the agenda.

The press release of 12 December after their second meeting declared that

[s]ince their foundation in the mid-1970s, the European People's Party and the Confederation of Social Democratic Parties in the European Community, and also the Federation of European Liberals and Democrats, have all in their own way made major contributions to European integration. Despite their political rivalry, and their opposed positions on numerous questions, both as regards content and method, all three European parties or federations of parties stress their common responsibility for the proper functioning of democracy and for the success of the European Union. To that end, they are working closely with their different political groups in the European Parliament. These groups play a major part in the continuing efforts to create a transnational consensus inside the different political families.[235]

Further meetings between the party leaders and Secretary Generals during 1991 served to prepare a formal initiative by the three chairs on 1 July 1991. On that day, a joint letter was sent to the presidents of the European

---

[234] Agenda for the meeting of 18 September 1990, EPP General Secretariat Archive.

[235] Press release, 12 December 1991, EPP General Secretariat Archive (Jansen's translation from French original).

Council, the Council of Ministers, the EP and the European Commission from the chairs of the three parties. The letter called for a clause on the role of European parties in shaping political consensus and political will to be written into the Treaty on European Union. The initiative for this letter came from the President of the EPP, Belgian Prime Minister Martens. It proved easy enough for him to convince both his colleagues and compatriots, Liberal De Clercq and Socialist Spitaels, of the significance of his proposal. This was intended, as the letter said, 'explicitly to emphasise the role of European parties in the process of integration and of democratising the European Union's political system'.[236]

The leaders of the three European parties proposed the following clause in the Treaty:

> European Parties are essential to integration within the Union. They are integral to building consensus and expressing the political will of the citizens of the Union. European parties are the federative associations of national parties with a presence in the majority of EU Member States, sharing the same aims and political direction, and forming a single group in the European Parliament. They must give a public account of where their funding comes from.

The first part of this formulation, inspired by Clause 21 of the German Constitution, makes clear that the role accorded to the parties in the context of the EU is essentially the same as that given to parties in a national context. In a federal state, parties and cross-party alliances essentially share the same function, and work along the same lines at their respective levels. But of course there will always be different ways of operating, depending on the competences to be taken into account at different political levels. It was in this spirit that the European party chairs felt that it would be possible 'to establish a European legal system in the medium term . . . which would give European parties a context for their work'.

Apart from their parliamentary groups, which operated on the basis of a statute sanctioned by the EP, not one of the legal preconditions for the activities of existing party confederations (or European parties) and their executive bodies was in place. That meant the parties had no legal status. They could not directly employ anyone, sign contracts or even give receipts for donations. To fulfil certain tasks, their secretariats had to pretend, both officially and in practice, to be acting on behalf of either their

---

[236] Letter from party leaders Martens, Spitaels and De Clercq, Brussels, 1 July 1991, EPP General Secretariat Archive.

group in the EP or individual member parties.[237] This was hardly a healthy state of affairs, and the practical effect was to severely impede the flexibility and capacity to react.

In the second part of the party leaders' proposal, European parties are described as 'federal alliances of national parties'. This is a reference to the associations representing the three classic party families – the Christian Democrats, the Social Democrats and the Liberals – whose members have committed to permanent cooperation based on agreed statutes and to programmes agreed upon by the appropriate bodies to put their common policies into practice. The goal of such cooperation in the federations is to create and maintain 'European unity of action by their members'. The EPP's rules of association further demand of member parties that 'they support the EPP's position in the European Community at national level. They further retain their names, identities, and freedom of action in the context of their national responsibilities.'[238]

The other elements of the definition set out in the draft article seemed important to its authors in order to set sensible legal parameters for party competences. A European party would hardly merit the name if it could not rely on national organisations in a number of Member States 'with the same direction and aims', and be able to unite the representatives of the various member parties in a single parliamentary group.

Reactions to the proposals by the three party leaders varied. Commission President Jacques Delors, with whom Martens discussed the matter, considered the initiative a positive one and encouraged it – *à titre personnel*. Its political importance was immediately clear to Delors, but he did not want to have the Commission involved, since it was a matter for parties and their parliamentary groups. So at the institutional level, it was initially a matter for the EP. This position was also confirmed by other members of the Commission who were approached on the matter.

The President of the EP, Spanish Social Democrat Enrique Baron Crespo, supported the party leaders' initiative and also took his own. Following a meeting on 2 October 1991 on how to proceed, he referred to the connection between the development of parties and the voting system and made the following suggestion: In addition to those elected on the basis of

---

[237] For instance, the EPP General Secretariat staff was for a long time partly employed by the EPP Group, and partly by the Belgian member parties.

[238] EPP General Secretariat Archive.

national voting systems, 82 deputies on European lists should be drawn up by the party confederations and elected. Voters would then have two votes, one for a national and another for a European list. The advantage of such a solution would above all consist in creating 'a purely European element' in the composition of Parliament, giving the European party federations a new and significant role. Moreover, the election of 82 extra deputies from European lists would improve the numerical relationship of voters to those they elected. Under this arrangement, the only decisive factor in the distribution of mandates would be the number of votes secured. Another effect of this proposal would be fairer representation for countries with large electorates, which were underrepresented as things stood. The whole idea promised a number of positive effects, one of them being greater commitment by Member States and their political forces to the European elections. It would also accelerate the process of political forces coming together at Union level, and strengthen party confederations.[239]

The reactions of Dutch Prime Minister Ruud Lubbers, President of the European Council, and of Foreign Minister Hans van den Broek, President of the EC Council of Ministers, were reticent, to a degree. Their attitude was influenced by severe disagreements within the IGC, which was preparing the Maastricht Treaty. In any event, the Dutch Presidency did not take up the suggestion. The draft Treaty proposed to the European Council on 9–10 December 1991 contained no reference to European parties.

The Maastricht Treaty: Article 138A

On 6 December 1991, the eve of the meeting of the European Council, an EPP Conference of Party and Government Leaders took place in The Hague. On the basis of a report by Lubbers there was a detailed discussion about the aims of the Maastricht Treaty and how far preparations had got.[240] Martens, in the chair, proposed a series of points that had remained open in the preparations for the IGC, and which the EPP regarded as essential. He put these to the meeting one after another, in each case making sure he had the

---

[239] Proposal by President Baron Crespo on uniform procedures for European elections, made at a meeting with Martens, Spitaels and De Clercq, Brussels, 2 October 1991, EPP General Secretariat Archive.

[240] Press Release, 12 December 1991, EPP General Secretariat Archive.

explicit backing of the other heads of government in the room.[241] Among the points was the issue of the draft article on European parties. Martens got the green light: all the Christian Democratic leaders agreed on the procedure and on insisting that the clause be in the text of the new Treaty.

There was no resistance to Martens's proposals at the European Council meeting. Only the Portuguese Prime Minister, Anibal Cavaco Silva, asked for clarification, and he was satisfied with the answer he received. The European Council then agreed without further intervention. But the wording of the article remained open, to be left to the conference of diplomats entrusted with editing the decisions and agreed texts of the Maastricht Summit. Article 138A of the Treaty on European Union, finally signed on 7 February 1992 by 12 foreign and finance ministers, reads as follows: 'Political parties at European level are important as factors for integration within the Union. They contribute to forming a European awareness and to expressing the political will of the citizens of the Union.' It is worth noting that the definition of 'European parties' contained in the party leaders' original proposal was not incorporated into the article, thus avoiding the laying down of a specific party model as the norm.

Article 138A of the Maastricht Treaty articulated the recognition that if the further unification of Europe were to be successful and a transnational system of government were to be effective, then the further development of European party structures would be necessary. At the same time, this constitutional recognition of the role and function of the parties served as an important basis for future efforts. The existence of 'political parties at European level' was eventually recognised. Parties were accorded the task of advancing the process of integration, of building a European consciousness and of expressing 'the will of the citizens of the Union'.

In practice, and mostly in anticipation of the coming into force of the new Treaty, Article 138A led to the (re)creation of the following European parties: the Confederation of Social Democratic Parties in the European Community, founded in 1974, rewrote its statutes and in 1992 became the Party of European Socialists (PES); the Federation of European Liberals and Democrats, founded in 1976, became the European Liberal Democrat and Reform Party (ELDR) in December 1993; the European Federation of

---

[241] Ruud Lubbers (the Netherlands), Helmut Kohl (Germany), Konstantin Mitsotakis (Greece), Guilio Andreotti (Italy) and Jacques Santer (Luxembourg).

Green Parties set itself up in the summer of 1993 as a pan-European association, but one whose framework was supposed to permit bringing these like-minded parties together at Union level; and the European Free Alliance/Democratic Party of European Peoples evolved in the second half of the 1990s into a federation of parties committed to ethnic and regional issues.[242] The EPP, having the intention of becoming a European party from its founding in 1976, even anticipated the outcome of the IGC by agreeing on new statutes in November 1990 that made this ambition crystal clear.[243]

Moreover, in the course of 1992, following the signing of the Maastricht Treaty, several meetings of the three party leaders took place. Also present were the chairs of the political groups and the President of the EP. Everything turned on the question of what was now to be done to breathe life into Article 138A and who was to do what. Two sets of problems loomed, inextricably bound up with each other, but perhaps needing to be dealt with separately: completing the picture with a 'law on parties' or a 'statute on parties', and the eventual possibility created by the new treaty of financing European parties with Community funds.

Establishing a Legal Basis for European Parties

It soon became clear to those taking part that until there was legal certainty, but also for reasons of political culture and morality, the financial question could not be posed until a number of conditions had been met. Unambiguous, legally binding rules about the organisation, activity and behaviour (including conduct of public finances) of European parties had to be put in place. But even leaving aside the question of financing the parties, it was felt that rules of this kind had become an urgent necessity. A statute on European parties would have to define what was meant by such concepts as 'European parties' and 'political parties at European level', what their tasks would be, what kind of rules would apply to their structures, and what kinds of working methods and finances they would have.

---

[242] Thomas Jansen, 'Zur Entwicklung Supranationaler Europäischer Parteien', in Oscar W. Gabriel et al. (eds.), *Der Demokratische Verfassungsstaat: Theorie, Geschichte, Probleme. Festschrift für Hans Buchheim* (Munich, 1992), 241ff.

[243] EPP Congress, Dublin, 15–16 November 1990. See also chapter six of the present volume.

This statute would have to define the essentials clearly enough for an independent, interinstitutional authority, or the European Court, to be able to identify a 'political party at European level' under Article 138A.

The EP's Institutional Affairs Committee felt these issues were its business and in May 1996 requested MEP Dimitros Tsatsos to prepare a report. It was ready by summer, and after debate and amendment in committee, was approved by a large majority in the December 1996 plenary session. The resolution accompanying the report called for directives about both the legal status and the financial circumstances of European parties.[244] This demand was in vain, for the time being at least. However, the report did encourage more rational debate in the EP among the political groups and parties, which contributed substantially to the agreement reached a few years later, when discussion about the future EU Constitution had become both livelier and more profound.

In December 2000, at the instigation of the European Commission, the EU heads of state and government decided, within the framework of the Nice IGC and on the basis of a Commission proposal, to introduce the following regulation as a clause in Article 191 (ex 138A), paragraph 2 of the EU Treaty, which could serve as the basis for agreement on a statute for European parties: 'In accordance with the procedures set out in Article 251, the Council is setting out the regulations for political parties at European level, and in particular rules about their financing.' An explanatory statement appended to the final act of the IGC stated that these regulations do not justify the transfer of any competences whatsoever to the EC, that they do not compromise the validity of rules deriving from national constitutions, that the use – direct or indirect – of funds from the Community budget for national political parties is prohibited, and finally, that the financial regulations should apply equally to all political forces represented in the EP.

The origin of this proposal was a joint effort by the transnational party families represented in the EP. On the sidelines of the European Council Summit in Feira, Portugal, in June 2000, they agreed to call for the following passage to be inserted into the Treaty: 'On a proposal by the Commission, the European Parliament and Council will decide on regulations on

---

[244] 'Report on the constitutional status of the European political parties' (known as the Tsatsos Report), Committee on Institutional Affairs, A4-0342/96; 'Resolution on the constitutional status of the European political parties', Official Journal C 020, 20/01/1997, 29.

recognising the political parties, on a statute for them, and on their financing.' The heads of state and government, it was generally supposed, had given their blessing to the party leaders' agreement, and so there were good grounds for supposing that the IGC would accept a proposal from the Commission along these lines.

Why exactly were the party leaders able to reach agreement and act together? They all found themselves in the same quandary: all of them had received a warning from the European Court of Auditors. The court had stated in a report that the parties were being financially supported, in part directly, from the parliamentary groups' budgets, that is, out of the budget of the EP, and that this was out of order and had to stop. This warning suddenly made the question of rules on European party finances very urgent. But first the status of the parties – that is, their legal position within the EU's political and institutional system – had to be resolved. Until the matter of their standing was decided, the question about party finances could not be resolved either.

By January 2001, the European Commission had already drafted European party statute and finance rules based on the new regulation in the Treaty. This provided for a budget line of seven million euros. However, the parties were not allowed to benefit from such funding unless all their financial dealings were subject, in their entirety, to the scrutiny of the Court of Auditors. This would apply even if only part of their revenues came from Community funds. The conditions were also set out for a party to qualify as European under the terms of the Treaty.

In May 2001, the EP agreed to a position based on a report by MEP Ursula Schleicher from the Bavarian CSU.[245] The proposed amendments were, for the most part, included in a Revised Draft Council Directive on a Statute and Financing for Political Parties. After the Council of Ministers failed to reach agreement, the European Commission proposed a Draft Directive by the European Parliament and the Council for a Statute and Financing for European Political Parties in February 2003. This took account of what had been discussed over the two previous years. Meanwhile, the Nice Treaty had come into force. This meant the co-decision procedure was applied now to Article 191, and the unanimity requirement for a deci-

---

[245] 'Report on the proposal for a Council regulation on the statute and financing of European political parties' (known as the Schleicher Report), Committee on Constitutional Affairs, A5-0167/2001.

sion by the Council fell away. Within only a few weeks, a working document report by rapporteur MEP Jo Leinen set out the arguments for the EP to consider when they debated the draft directive. At the June plenary session, the statute for European parties was duly adopted; the Council of Ministers agreed on the text at the same time.[246]

The 2004 Regulation

The 2004 regulation – in full, Regulation (EC) No 2004/2003 of the European Parliament and of the Council of 4 November 2003 on the regulations governing political parties at European level and the rules regarding their funding – came into force on 15 February 2004. What does it say? The legislative act sets four conditions for a 'political party at European level'. First, 'it must have legal personality in the Member State in which its seat is located.' Second, 'it must be represented in at least one quarter of Member States, by Members of the European Parliament or in the national Parliaments or regional Parliaments or in the regional assemblies, or it must have received, in at least one quarter of the Member States, at least 3% of the votes cast in each of those Member States at the most recent European Parliament elections.' Third, 'it must observe, in particular in its programme and in its activities, the principles on which the European Union is founded, namely the principles of liberty, democracy, respect for human rights and fundamental freedoms, and the rule of law.' Fourth, 'it must have participated in elections to the European Parliament, or have expressed the intention to do so.'[247]

'In order to receive funding from the general budget of the European Union, a political party at European level must file an application with the European Parliament each year. The European Parliament shall adopt a decision within three months and authorise and manage the corresponding

---

[246] Proposal KOM (2003) 77; Working Document of the Committee on Institutional Affairs–DT\488199DE.doc; Directive (EG) No. 2004/2003 of the European Parliament and Council of 4 November 2003, on regulations governing political parties at European level.

[247] Summary of Procedure File 'Political parties at European level: statute and financing', Legislative Observatory (http://www.europarl.europa.eu/oeil/file.jsp ?id=5505072).

appropriations.'[248] The regulation stipulates strict rules with regard to donations and financial dealings. The EP is responsible for ensuring that the rules governing eligibility for financing are observed, and has the right to disqualify from receiving further funding parties that disobey the rules. Furthermore, the Court of Auditors can inspect any document or source of information related to the parties' activities. The regulation also says that

> [c]ontributions from political parties which are members of a political party at European level shall be admissible. They may not exceed 40% of that party's annual budget. The funding of political parties at European level from the general budget of the European Union or from any other source may not be used for the direct or indirect funding of other political parties, and in particular national political parties, which shall continue to be governed by national rules.[249]

How much money will the 'political parties at European level' get? '[Fifteen percent] shall be distributed in equal shares and 85% shall be distributed among those which have elected members in the European Parliament, in proportion to the number of elected members . . . Funding charged to the general budget of the European Union shall not exceed 75% of the budget of a political party at European level.' In practice, the EP recognised eight parties in 2004 and distributed 3.22 million euros among them, including the 1.05 million euros received by the EPP.[250] Since then the overall amount of money to be distributed has been augmented and the number of parties that are recognised as 'political parties at European level' has risen to 10.[251]

---

[248] Ibid. See 'Decision of the Bureau of the European Parliament of 29 March 2004 laying down the procedures for implementing Regulation (EC) No 2004/2003 of the European Parliament and of the Council on the regulations governing political parties at European level and the rules regarding their funding', OJ 2008/C 252/01.

[249] Summary of Procedure File 'Political parties at European level: statute and financing', Legislative Observatory (European Parliament website.)

[250] See http://www.europarl.europa.eu/pdf/grants/grant_amounts_parties.pdf.

[251] Ibid. All relevant documents, including amounts of grants, can be consulted at the website of the European Parliament, www.europarl.europa.eu. See Grant to political parties and foundations.

The 2007 Amendment

Despite the relief of finally having established a solid basis for European political parties, at least financially (as the question of statute was not fully answered yet), the idea soon arose of improving the rules governing the financing of political parties. The legislative process took place in the second half of 2007, resulting in Regulation (EC) No 1524/2007 of the European Parliament and of the Council of 18 December 2007 amending Regulation (EC) No 2004/2003 on the regulations governing political parties at European level and the rules regarding their funding. The Commission laid down a proposal at the end of June 2007, the report by MEP Leinen was adopted by the EP at the end of November and the Council approved the document in first reading a few weeks later.

What did the 2007 regulation change? First, its purpose was 'to strengthen the potential of long-term financial planning of political groups and to facilitate the diversification of financial resources. It aims to grant greater flexibility to political parties ahead of the parliamentary elections taking place in June 2009.' Therefore,

[t]he expenditure of political parties at European level may also include financing campaigns conducted by the political parties at European level in the context of the elections to the European Parliament, in which they participate. These appropriations shall not be used for the direct or indirect funding of national political parties or candidates. Such expenditure shall not be used to finance referenda campaigns. However, the funding of and limitation of election expenses for all parties and candidates at European Parliament elections is governed in each Member State by national provision.[252]

At the same time, the maximum level of the political parties' budget that is granted by the EP was raised from 75 to 85%.

Second, grants of the EP were also given to so-called 'political foundations at European level', defined as 'an entity or network of entities which has legal personality in a Member State, is affiliated with a political party at European level, and which, through its activities, within the aims and

---

[252] Summary of Procedure File 'Regulations governing political parties at European level and rules regarding their funding (amend. Regulation (EC) No 2004/2003)', Legislative Observatory (European Parliament website.)

fundamental values pursued by the European Union, underpins and complements the objectives of the political party at European level'.[253]

Political foundations received grants for the first time in 2008.[254] In 2009, the first full calendar year in which the new regulation applied, about 6.7 million euros was set aside to be distributed among nine political foundations. The EPP's political foundation, the Centre for European Studies, received almost 2.3 million euros.[255]

Towards Truly European Party Organisations?

How should one evaluate this 2004 regulation and its 2007 amendment? The new rules give European political parties financial stability and a clear framework in which they can develop their organisations and activities. Funding is now more transparent, and the rise in amounts granted by the EP has facilitated the parties' development and, paradoxically and perhaps more importantly, their independence from the EP's party groups.[256] One should, however, also add that a lot of the costs (such as translation services during meetings of the EP) that previously were paid by the party groups are now paid by the parties themselves. The increased budget also triggered professionalisation as far as infrastructure, personnel and communication were concerned. For the EPP this meant a new party headquarters building (which increased the EPP's street visibility in Brussels' European quarter), hiring more people from different member parties and developing the party's website and electronic communication.[257]

The extension of the original regulation towards political foundations was clearly a novelty, compared with the agenda with which the European political parties started in the early 1990s. For the first time European political parties were given the potential to establish research departments that would primarily focus not on forthcoming EP elections but on strengthening the parties in their long-term development – not as organisa-

---

[253] Ibid. See chapter twelve of the present volume.
[254] From October 2007 to August 2008 grants were awarded by the European Commission under a pilot project.
[255] See http://www.europarl.europa.eu/pdf/grants/grant_amounts_foundations.pdf.
[256] See chapter thirteen of the present volume.
[257] Interview with Luc Vandeputte, EPP Deputy Secretary General, Brussels, 10 February 2010.

tions or institutions but as political families – generating ideas about the future of the EU, its institutions and above all its policies, and European societies. The underlying message is certainly that parties have a role to play beyond the day-to-day politics of EU Brussels. Establishing political foundations also offers the opportunity to strengthen relationships with its member parties, since their national political foundations give body and soul to the European political foundations.

Many things, however, still need to be done. The most important item outstanding from the 2004 regulation and the 2007 amendment is a clear statute for European political parties, a statute that is more than the obligation to have legal personality. Because of the 2004 regulation, the EPP, for instance, turned its statute into an *Association internationale sans but lucratif* (a.i.s.b.l.), that is, an international non-profit organisation, under Belgian law. A legal identity of this kind is a precondition for the EPP – as for the other European parties – to claim and manage the funds made available from the EU budget under the provisions of party regulation. This is an unusual legal identity for a political party, all the more so in that it forces certain formalistic acrobatics. That in turn is a clear sign that all this can only be provisional, something which needed to be devised because both the legislative bodies – the European Council and Parliament – had so far either failed or were not in a position to endow the European parties with European legal personalities within the framework of the new statute. This matter, together with other issues such as special funding for European election campaigns and more stringent recognition criteria, will need to be dealt with in the near future.

Looking back at 20 years of fighting for the recognition, in principle and in practice, of the role of European political parties, it is clear that the rhythm of their development was in fact largely something they determined themselves. After all, the EU is run by politicians, even though some of the representatives now and then tend to forget to which political family they belong and the extent to which they are indebted to that family. Their development is also due to the importance the Union has attributed to democracy, transparency and closeness to the citizen. If anything, European political parties should 'contribute to forming a European awareness and to expressing the political will of the citizens of the Union'. It is therefore no surprise that the Lisbon Treaty (as well as the failed Constitutional Treaty) folded the former Article 138A into a larger Article 10 that deals with aspects of representative democracy (under the heading Title II: Provisions on Democratic Principles).[258]

Playing their role within an emerging European representative democracy, the European parties are neither able nor willing to copy any particular national model or its inherent political culture. They must respect their member parties' existing mature and proven structures, building on them and depending on them. In other words, they are becoming 'federal parties' seeking to organise the joint activities of their members at European level and ensure that their efforts are politically effective. This is anything but easy, since what a European political party wants to become does not automatically correspond to what individual member parties would like to make of it. Ideas about what a European party should be or achieve will vary considerably between member parties (and between EU Member States). People normally orient themselves by what they know – their own national party, and whatever European or transnational sensibility they happen to possess. For European political parties, it is crucial to take this into account.

In other words, European political parties develop in an open force field and are subject to a kaleidoscope of different influences. Elements of all the different kinds of national parties will leave a trace in the European parties, whose essence must, eventually, be something different. Yet to a greater or lesser extent, member parties expect the European parties to conform to the preconceptions they have brought from home. There is often an inclination to adjust the parent party's image and achievements to domestic criteria. This explains the tendency to exploit the European parties to advance national party interests, or to measure the value of the European parties by their direct usefulness in particular situations. These reflexes are typical during the transition to a new political system, a phase when new behaviours are still unfamiliar and old experiences remain the model. Given the nature and the direction of recent changes, however, it is likely that in such a new political system, more than ever, European political parties will be of central importance.

---

[258] 'Consolidated versions of the Treaty on European Union and the Treaty on the Functioning of the European Union', Official Journal C 115 of 9 May 2008.

Chapter Twelve

The Centre for European Studies[259]

From the beginning of the creation of the EPP, I was convinced that we also needed European political foundations. With the creation of the EPP in 1976, I drafted the first by-laws for the foundation of the EPP. However, political parties had not yet obtained the financial means to fund these foundations. This process would have to wait almost 30 years. But finally, European political parties were in a position to discuss the establishment of European political foundations.[260]

The daily work of national political parties is fast paced, focused on developments and debates in the media and national politics. Party officials are required to react quickly, and the daily televised newscasts and deadlines set by the press determine the pace of the party's life. But this intense rhythm comes with a price. There is a tendency to favour fast reactions and political repositioning over the long-term planning and reflection necessary for a political party to maintain its cohesion and define political positions in the changing world. Even if long-term reflection is often envisaged, the party machinery has to prioritise, and long-term plans and reflections come second to the immediate challenge of achieving victory in elections. Thus, the process of deeper study and reflection is interrupted and can even be forgotten, at least for a while. In the long run, a lack of accumulation of knowledge and analytical reflection within political parties will have a negative impact on the quality of their political positions. Therefore, in order to ensure more long-term political work, national parties often establish political institutions, research and training centres, and political foundations as separate but affiliated organisations.

---

[259] The author, Tomi Huhtanen, would like to thank Elaine Larsen, Boyan Tanev and Lauren Zoebelein for their research assistance with preparing this chapter.

[260] Interview with Wilfried Martens, 9 November 2009.

European political foundations can be traced back to similar German institutions launched after the Second World War to support the development of the newly established political system. Each party represented in parliament would have guaranteed public financing, depending on its proportion of seats. The term 'political foundation' can be misleading, since a foundation is normally understood to be a legal entity, a non-profit organisation that will either donate funds and support to other organisations, or provide funding for its own purposes. Instead, political foundations are non-governmental organisations connected to political parties and often supported by public funds. Their mission is to foster progress in society by contributing to political education, information and debate. In practice, this broad definition covers a wide array of groups with different activities, and in many countries a legal framework for political foundations does not exist. Political foundations range in size from large organisations like the German Stiftungen with extensive networks of offices around the globe to small organisations with only a few staff members, labelled as institutions, study centres or political foundations, depending on the political culture and the size of the country. This type of organisation in Europe may have very different focuses of activity, some of them concentrating their efforts on training, others on organising political events, or on publications and promoting democracy. Some of them work very closely with the political party with which they are affiliated, and many of them aim to engage in independent political reflection, working as think tanks.

The idea of establishing European political foundations existed from the 1970s and was first presented by former Prime Minister of Belgium and European People's Party (EPP) President Wilfried Martens. However, the real implementation of the idea took place only when the legal framework of European political parties was established three decades later. Incorporating the existing model of the political foundations into the framework of European Union (EU) legislation was a complicated process involving various stages. The provisions for the existence of European political foundations were incorporated into an existing regulation that provides a framework for funding European political parties.[261] Originally created in

---

[261] Regulation (EC) No. 1524/2007 of the European Parliament and the Council of 18 December 2007 amending Regulation (EC) No. 2004/2003 on the regulations governing political parties at European level and the rules regarding their funding, Official Journal of the European Union, L343/5-8, 27 December 2007.

2003, the legislation for political parties was to be reviewed in 2006. It was at this point that the idea occurred to amend legislation to also incorporate funding for European political foundations. Their creation is part of a larger policy strategy to promote European ideas and dialogue to the greater European public, ideas that easily aligned with the European Commission's efforts to fulfil a policy strategy referred to as 'Plan D': Democracy, Dialogue and Debate.[262] The goals of Plan D were to lay the foundations for a debate about Europe's future, in the context of the French and Dutch 'no' votes on the European Constitution, and to build a political consensus on the challenges for the EU.

Thus, the establishment of these foundations occurred at a crucial time and in a fragile political climate. In 2005, the constitutional referenda in France and the Netherlands exposed a sizeable and growing disenchantment among Europeans on the subject of further EU integration. Moreover, it was understood that inaccurate information was being disseminated to the public by Euro-sceptic movements, distorting fundamental European issues. That proved it was imperative for the Commission to produce a policy initiative that would establish an accurate dialogue with the public. At the same time, an examination by the Commission revealed holes in the infrastructure of national think tanks and political foundations across the EU. Generally speaking, older Member States had stronger political foundations, in stark contrast to the weaker traditions, and therefore weaker foundations, of newer Member States.

Together with the Commission, policymakers from the European Parliament (EP) and the Council of Ministers collaborated to establish European political foundations, whose purpose was to 'promote greater understanding, debate, and new thinking, as well as to constitute a channel through which a greater number of citizens can actively participate in European democracy'.[263] European political foundations were seen as an influential tool in providing a long-term solution to this problem.

The launching of the European political foundations had two phases. The first was the 'Pilot project' phase, which lasted from 1 September 2007 to the end of August 2008, and enabled the European foundations to start

---

[262] European Commission, 'European Commission launches PLAN D for Democracy, Dialogue and Debate', Brussels, 13 October 2005, IP/05/1272, http://europa.eu/rapid/pressReleasesAction.do?reference=IP/05/1272 a.

[263] European Commission, Call for Proposals DG EAC/29/2007, European political foundations: Pilot projects, 3.

with limited funding. However, during that time the European political foundations had no status in European legislation. Official status was obtained on 27 December 2007, which marked the beginning of the second phase and made it possible for European political foundations to obtain grants from the EP after the Pilot project phase was finished in August 2008.

Towards a Parliamentary Initiative

In March 2006 the EP began to actively advocate for greater flexibility for the financing of political parties, most notably for the creation of legislation for political foundations at the European level. The foundations would be institutions intended to have greater value than those at national level and would provide all Member States with a common European framework to facilitate public discussion. The proposed political foundations would also seek to improve the democratic quality of European structures.

All mainstream European political parties and their affiliated groups in the EP, apart from the Euro-sceptics, were unanimous in their position that the creation of European political foundations would provide a genuine boost to dialogue on European current affairs. Under the co-decision procedure, the Parliament used the budgetary aspect of the proposed regulations to approach the Commission with an 'own-initiative report'. The Commission then prepared the text for a legislative proposal, which was submitted on 27 July 2007, after a remarkably short amount of time. Exhaustive and contentious negotiations followed, however, before the legislation was eventually adopted.

When the proposal was returned to Parliament by the Council on 22 October 2007, three characteristics of the draft provisions, which were considered by Parliament to be essential components of the proposal, were questioned by the Council before its final reading. First, there was doubt as to whether the relationship between a political foundation and its affiliated party should be kept vague. The final agreement was that this would stay so, proclaiming that 'the activities of political foundations at the European level underpin and complement the objectives of the political party at the European level'.[264] Second, an interesting debate developed between old

---

[264] Proposal for a Regulation of the European Parliament and the Council amending Regulation (EC) No 2004/2003 on the regulations governing political par-

and new Member States on the interactions of these proposed new political foundations with third countries. New Member States, especially those from Central Europe, were in favour of having the proposed European political foundations explore EU–third country issues, partly as a tool to spread European values to these countries. After later amendments of the Council to regulate their activities outside of the EU Member States, one of the defined tasks of the European political foundations became to develop cooperation with entities of the same kind in order to promote democracy. A third discussion involved the role of political foundations in the campaign strategies of their affiliated parties. It became clear that it was not appropriate to have political foundations involved in campaigning, and it was decided that they would remain neutral, but play a role in making citizens aware of issues relating to European elections.

Cooperation with the Commission

The European Commission was at first dubious about the proposal advocating greater flexibility for the financing of political parties. Fortunately, it was a high priority for the Commission's President, José Manuel Barroso, and consequently for the Commission's Vice-President, Margot Wallström, responsible for Institutional Relations and Communication Strategy, to nurture positive relations with the EP, and to commit to the above-mentioned 'Plan D' initiative. Commissioner Wallström simultaneously began making preparations to execute the Pilot project phase, funded by the European Commission.

Political affiliation to the various parties was thought to be a suitable arrangement for the foundations. It provided that the foundations would comply with the same financial model and rules as those of European parties so that to ensure the financing of European political foundations, the parties had to assume some degree of responsibility for them. In the end, a proper balance regarding affiliation, party responsibility and foundation autonomy was reached, and Article 191 concerning European political parties was extended to also cover European political foundations.[265]

---

ties at European level and the rules regarding their funding, COM (2007) 364 final, Brussels, 27 June 2007, 4.
[265] See chapter eleven of the present volume.

According to the regular procedure, the Parliament was to vote on the creation of a Pilot project, and then after a year or two, a programme evaluation would decide whether it was worthwhile making the project permanent. This situation was atypical, however. The Pilot project and the creation of the regulations proceeded simultaneously as a result of a strong political will to move quickly with the legislation. The text for the proposal was adopted by the Commission on 27 June 2007. The Parliament ensured the proposal would be created and adopted as quickly as possible so that parties could benefit from the budgetary amendments of the 2008 fiscal year. This would also allow for the European political foundations to be operative in preparation for the upcoming European elections in 2009.

Finally, after various amendments, a 'European political foundation' was defined as

an entity or network of entities which has legal personality in a Member State, is affiliated with a political party at European level, and which through its activities, within the aims and fundamental values pursued by the European Union, under-pins and complements the objectives of the political party at European level by performing, in particular, the following tasks:

- observing, analysing and contributing to the debate on European public policy issues and on the process of European integration;
- developing activities linked to European public policy issues, such as organi-sing and supporting seminars, training, conferences and studies on such issues between relevant stakeholders, including youth organisations and other repre-sentatives of civil society;
- developing cooperation with entities of the same kind in order to promote de-mocracy;
- serving as a framework for national political foundations, academics, and other relevant actors to work together at European level.[266]

Council Commitment

After the text of the proposal was created, the Parliament and the Commis-sion managed to forge an agreement with the Portuguese Presidency to

---

[266] Regulation (EC) No. 1524/2007 of the European Parliament and of the Council of 18 December 2007 amending Regulation (EC) No. 2004/2003 on the regu-lations governing political parties at European level and the rules regarding their funding.

speed up the Council negotiations. The President handed the proposal for assessment to the Council Working Group for General Affairs. This was a political strategy in itself, because had it been handed to the Working Group for Budget, the proposal would easily have been terminated by those Member States who were not in favour of it. Yet, many Council members were concerned about the approval of funding for entities that did not exist, and many wished to proceed through negotiations with caution.

Though the establishment of European political foundations was intended to benefit specifically newer Member States, surprisingly, the proposed regulations worried Council members primarily from those countries. Many of the newer members were not familiar with the concept of political foundations, and initially contended that they did not see the benefit of establishing them in their countries, which did not have an existing network of European-oriented organisations, think tanks, foundations and so forth that could be used as an operational network. The issue was investigated and overcome through research conducted by European political parties, which determined that the presence of political foundations linked to European parties would help to strengthen existing national European frameworks. Thus, national member foundations could exist as a part of the organisation of the European foundations. However, some smaller EU Member States remained concerned that the European foundations would be quickly dominated by the representatives of larger EU Member States, which had a strong, established culture of political foundations. Therefore, in the regulation it was stipulated that the governing body of a European political foundation would have a geographically balanced composition.

The lobbying efforts of the Parliament and the strong statement of political support by the President of the EP, Hans-Gert Pöttering, proved to have significant influence in upholding the commitment to keep the proposal alive and work through all the technical issues. The entire proposal was adopted on 27 December 2007, just in time to be incorporated into the 2008 budget. Six months was all it took for the normal co-decision procedure to be implemented; it was one of the shortest legislative procedures ever carried out.

The Pilot Project

The European Commission directed and monitored the Pilot project for European foundations under the Directorate-General of Education and Culture, within the programme 'Europe for Citizens', which had a mandate complementary to that of the political foundation Pilot project. This way, the two projects could be run in tandem. Meanwhile, negotiations for the legalisation of political foundations were taking place in Parliament. This was unconventional; the usual process dictated that the results of a pilot project be evaluated, and only when the project had been approved on the basis of its performance would a motion for permanent action be registered.

All of the foundations were created between the summer and autumn of 2007 and became registered as legal personalities under Belgian law. The launching of the political foundation of the EPP, the Centre for European Studies (CES), was accomplished with the cooperation of EPP officials. EPP President Martens and EPP Secretary General Antonio López-Istúriz had a decisive role and devoted considerable time to the cause from 2006 to 2008. Deputy Secretary Generals Christian Kremer and Luc Vandeputte actively negotiated with the Commission and helped formulate the framework for the European foundations. EPP External Relations Secretary Kostas Sasmatzoglou (later EPP Spokesman), was involved in the strategic planning and architecture of the EPP foundation, and helped in influencing the decision-makers in the EP and the Council of Ministers. For the EPP and the newly founded CES led by Director Tomi Huhtanen, the challenge was to simultaneously manage the rapid launching of the CES, to administer its activities, and to influence the Commission, the Parliament and the Council in order to ensure that the legal framework would be fully set in place.

Prior to the advent of the CES, the EPP had organised conferences that carried a likeness to the activities of political foundations. *European View* was an academic periodical already being published by the EPP. The know-how existed for conducting such activities. The EPP already had networks and organisations in place that operated as think tanks or training centres for the EPP family, including the European Ideas Network of the EPP Group in the EP, and the Robert Schuman Institute, a democracy-building and training organisation that cooperated with both the party and the Group.[267]

It was agreed among all the actors in the EPP family that the CES would function exclusively as the party's foundation and think tank. In the debate on the relation of the EPP foundation to the already existing national political foundations or research and training institutes, the representatives of the national political foundations had two basic concerns. First, they considered it unnecessary to have a European umbrella organisation representing them vis-à-vis the institutions of the EU. Furthermore, many of the national member foundations considered it unnecessary for the CES to focus on democracy promotion, since various national foundations had a long tradition in this field. There were also fears that European and national foundations would compete for the same resources to promote democracy. It was agreed that the EPP foundation would focus primarily on think tank work, such as research, analysis, observation, organising seminars and conferences, and supporting the networking of national foundations.

A major question both for the EPP and the national foundations was the separation of the party and its foundation. For the national foundations, following the tradition of many Member States would have meant that the party and the foundation were strictly separated from each other. However, this was not the case in the European legislation. Since the European foundations did not have a separate regulation but were defined in the same regulation as the parties, the challenge for the legislator in assessing the impact of the regulation was not too much closeness but too much separation. This in fact was the Commission's concern in the beginning while drafting the proposal. As a consequence, some of the ideas for the future structure of the EPP foundation included a model whereby the EPP Presidency and foundation governing board completely overlapped. The Commission never took an official position on this governing model, but signalled that such a construction would not be politically ideal. The final approved regulation expressed that:

> Within the framework of this Regulation, it remains for each political party and foundation at European level to define the specific modalities for their relationship, in accordance with national law, including an appropriate degree of separation between the daily management as well as the governing structures of the political

---

[267] See chapter four of the present volume.

foundation at European level, on the one hand, and the political party at European level with which the former is affiliated, on the other hand.[268]

This was a formulation which gave parties and foundations the opportunity to define their relation in a flexible manner. However, it was later confirmed by Parliament Bureau decisions that the funding of the foundations was accessory to the funding of the European parties.[269] In addition, the funding of the foundations depended on the yearly grant applications submitted by the foundations themselves, and was always conditional on the confirmation of its respective European political party that the foundation in question was really entitled to receive the funding from the EP to run its activities as a European political foundation affiliated to that European party.

With regard to the EPP, the common vision of the model of the foundation was drafted in late spring 2007 and the main principal accord between the representatives of the EPP and the national foundations was reached in August 2007 in Muskö, near Stockholm. Of the many representatives of the national foundations the key personalities were Peter R. Weilemann, Director of the Konrad Adenauer Foundation Brussels office; Eva Gustavsson, Managing Director of the Jarl Hjalmarson Foundation; Susanne Luther, Head of the Office for Foreign Relations of the Hanns Seidel Foundation; and Erik Kroiher, Head of the International Office of the Political Academy of the ÖVP.

The CES is defined by two pillars – the party and the member foundations – in an equal relationship. The model agreed was a structure in which the governing bodies, the General Assembly and Executive Board, consists of representatives of the EPP and the national foundations – half from each constituent group. The Executive Board consists of eight members: three are proposed by the EPP Presidency, including the foundation's President, its Secretary and its Treasurer; three represent the national foundations and two are individuals selected by common agreement. The General Assembly is composed of organisations affiliated to EPP member parties with a

---

[268] Regulation (EC) No. 1524/2007 of the European Parliament and of the Council of 18 December 2007 amending Regulation (EC) No 2004/2003 on the regulations governing political parties at European level and the rules regarding their funding.

[269] European Parliament Bureau Decision of 17 December 2007 on Financing of Political parties and Foundations.

limit of one organisation per party. The EPP is represented by individual members proposed by the EPP Presidency. This model, on the one hand, vested the EPP foundation with expertise and an extensive network of solid national partner foundations, and, on the other, offered political support and high-level national political contact from the EPP and its member parties.

Therefore, the stage was set for the founding of the CES. On 13 September 2007, after the first day of a regular EPP Political Bureau meeting, the representatives of the EPP and national foundations gathered to sign the founding documents. The founders of the CES were EPP President Martens, EPP Secretary General López-Istúriz, and eight national foundations, namely the German Konrad Adenauer Foundation represented by Weilemann, the Greek Constantinos Karamanlis Institute for Democracy represented by Pantelis Skilas, the Spanish Foundation for Social Research and Analysis (Fundación para el Análisis y los Estudios Sociales, FAES) represented by Juan Magaz van Nes, the Bavarian Hanns Seidel Foundation represented by Luther, the Swedish Jarl Hjalmarson Foundation represented by Gustavsson, the Austrian Political Academy of the ÖVP represented by Kroiher, the Hungarian Foundation for a Civic Hungary (Polgári Magyarországért Alapitvány) represented by Zoltán Balog and the Dutch Scientific Institute of the CDA represented by Jan Evert van Asselt.

'Running Start' of the CES

The Pilot project for the CES was led by Director Huhtanen and Project Manager Nicholas Alexandris – who became the first employee of the think tank – and consisted of common projects with all eight founding member organisations.

In January 2008 the first CES General Assembly took place and the Executive Board was selected. It included EPP President Martens, EPP Secretary General López-Istúriz, Joseph Daul, János Martonyi, Yannis Valinakis, and the Konrad Adenauer Foundation represented by Weilemann, the Scientific Institute of the CDA represented by Raymond Gradus and the Jarl Hjalmarson Foundation represented by Margaretha af Ugglas. The Executive Board also nominated Werner Fasslabend as first Chair of the Academic Council. The first individual members elected in the following General Assembly meetings included Klaus Welle, Miguel

Papi-Boucher, Charles-Ferdinand Nothomb, Kostas Sasmatzoglou, Filippo Terruso, João Marques de Almeida, Kristof Altusz, Antti Timonen, Jacob Lund Nielsen, Nicolas Pascual de la Parte, Timothy Beyer Helm and Marianne Thyssen.

In autumn of 2008, the CES had been operating for four months and had secured new employees and new offices. With these preparations it was able to launch its 2009 agenda. The Centre presented an International Visitors Programme in Warsaw to coincide with the EPP Congress, beginning the Centre's first engagement with business stakeholders. The CES then created an EU Elections Watch series of electronic papers, which provided chronological pre- and post-election analysis of the successes of the European elections in each Member State. It also launched the 'tellBarroso.eu' project, which provided a Web-based platform with the participation of 150,000 European citizens to contribute to the political decision making, an initiative personally endorsed by Commission President Barroso. After the launch of these initiatives in early 2009, it was clear the CES had made the jump from newcomer to player in the Brussels think tank world.

Beginning in early 2009, the CES implemented a research plan of eight to nine major projects for completion in 2009 and early 2010. Since the CES does not have a large staff, but rather a large network through its member foundations, its strategy for conducting research has been to co-operate with external researchers linked to CES member foundations and the EPP member parties. The work consisted of research papers, strategy papers and policy briefs, but also larger publications. CES research aims to follow the topics relevant to the EPP political debate and has also tackled six different areas of challenge, both for the EPP and for Europe: Institutions and Process of Policy; Economic and Social Policy Reforms; EU Foreign Policy; Environment and Energy; Ethics, Values and Religion; and European National and Regional Policies.[270] The first CES Head of Research and Deputy Director was Roland Freudenstein. As a result of the regulatory obligation to secure 15% of its budget from its own resources, the CES began fundraising activities, which were launched in the midst of the European financial and economic crisis, by Communication and Marketing Manager José Luis Fontalba.

---

[270] See http://www.thinkingeurope.eu/index.php.

The CES has maintained a balanced relationship with the EPP, which has allowed the think tank to create compelling products with a clear message, and at the same time to provide for a strong voice of the centre-right in the Member States. In the future it will remain necessary to strike a good balance between the thematic interests of its work and the research demands of the CES member foundations and stakeholders of the EPP.

Chapter Thirteen

The EPP's Relationship to the Group in the European Parliament

The relationship of the party to the Group in the European Parliament (EP) has been central to the founding and development of the European People's Party (EPP). For a long time, the party heavily depended on the Group with regard to both resources and exposure: there was no party outside the Group.[271] Changes at various levels have altered this situation. Parties and groups now behave in a much more 'adult' way. They need to work together, particularly prior to and during EP elections, but some tension between them will continue to exist.

Looking for a Balance

In most parliamentary democracies with active political parties there is invariably a palpable antagonism between the parties and their parliamentary groups. How charged such tension becomes depends on a variety of factors, including the personalities involved and the political situation of the parliamentary group – whether the party is in power or opposition. The overall institutional picture and the political culture are both relevant as well. Normally, a strong parliamentary group will have a strong party behind it, since the group is an expression of the party, and the party will be, by the same token, to a large extent shaped by the group.

---

[271] See for instance, 'The PPE [sic] Group at the European Parliament is not only a reflection of the unity of European Christian Democrats. It is the initiator, the guide and the spearhead in the attainment of our objectives at Community level'. Egon Klepsch in *EPP Bulletin: Communications from the General Secretariat of the European People's Party* 6, (September/October 1986), 6.

A parallel tension exists between parties and their groups at European level, but in this case the shoe is very much on the foot of the parliamentary groups. Matters remain as they have from the beginning: European parties play no role at all in choosing candidates for the European elections. As long as the right to vote in European elections remains a national right, it will be the preserve of national parties in the Member States to pick the candidates and organise the campaigns. The organisational weakness of the European parties also needs to be taken into account. Until their legal position in the Union was resolved, as happened only recently, they remained financially dependent on their component parties. Such dependence was not in itself problematic, since the European parties are ultimately made up of national parties. But given the national parties' notorious reluctance to make available the finances necessary for effective work by the European parties, there was only one solution to their funding problem.

This solution was to turn to the groups in the EP. And these – both out of self-interest and because they share the same European mission – are invariably more sympathetic and willing to support the European parties. At an early stage, when the national parties scarcely recognised the need to do so, European parliamentary groups made available the resources and structures necessary to take the first steps towards uniting the parties as single organisations. The contribution of the parliamentary groups, both financial and otherwise, was from the start considerably greater than that of individual member parties – that is, until the Court of Auditors raised objections.

The special role of parliamentary groups in, and in relation to, the European parties is also a function of history. From the outset, parliamentary groups brought together political figures with the requisite knowledge of how Europe worked. These were the people who convinced party leaders at home of the importance of closer cooperation with partners in neighbouring countries. And it was they who took the initiative to establish European parties. Those parties were, in the first instance, the offspring of the parliamentary groups of the EP, which meant that from the start, the groups had a strong influence on their progeny. As co-founders, the parliamentary groups are also – in addition to the member parties – constituent members of the European parties. One effect of this is the powerful role reserved for the groups and their representatives in the party statutes.

Indeed, the origins and development of the EPP cannot fully be understood without taking into account its Group in the EP. The leadership of what was originally the Christian Democratic Group played a crucial role in the establishment of the party.[272] In those years and during a long period thereafter, the EPP was heavily dependent on the Group, especially in financial and organisational matters. At the same time, and irrespective of its own development, the Group gradually became linked with the party.[273] There was no longer one without the other. This was already clearly illustrated by the Group's name change in 1999. The most explicit and important collaboration between the EPP and its Group is of course the EP elections every five years, particularly the campaign.[274]

European election campaigns are invariably, in every single Member State, exploited by national party leaderships for their own immediate purposes. This leaves little room for European parties to develop a political presence. They are essentially invisible as actors, and can therefore not have a profile. It is true that programmes jointly developed and agreed between the national and federated parties of each of the European party federations are made available to the campaigning member parties, but little is done consistently to 'sell' European policies.

The 1979 Elections

In the first European campaign, member parties to varying degrees tried to appear cohesive, referring to the EPP platform and the common political programme that had been adopted. The EPP leadership could exercise a certain influence as far as uniform presentation was concerned in individual countries, thanks to the financial resources the EP put at the disposal of parliamentary groups or their political forces. These resources were intended to pay for disseminating information and publicity about the direct elections. The Political Assembly, then Political Bureau, and the General Secretariat tried to coordinate the publicity campaigns, and some events

---

[272] See chapter two of the present volume.

[273] For a very extensive and detailed history of the Group, see Pascal Fontaine, *Voyage to the Heart of Europe 1953–2009: A History of the Christian-Democratic Group and the Group of the European People's Party in the European Parliament*, (Brussels, 2009).

[274] The party programmes are equally important but are, in the framework of the electoral Congresses, dealt with in chapter six of the present volume.

with a European flavour featuring well-known politicians of different nationalities were sponsored jointly with national parties.

EPP parties were very successful in the first direct European elections. Altogether they attracted 32.6 million votes, or 29.6% of the electorate. This was a better result than that of the Socialists, who managed only 29.3 million votes (26.6%). It was a remarkable leap forward, especially given that the EPP was not represented in the United Kingdom and Denmark, and the Social Democrats had member parties throughout all Member States. The divergent national electoral laws and the fact that obtaining a mandate required very different numbers of votes in different Member States had the curious effect of putting the EPP Group and its 107 deputies in second place in the EP – right behind the Group of the Social Democrats, which had 112 members.

The 1984 Elections

In June 1984, ahead of the second European elections, the EPP Political Bureau and General Secretariat again attempted to coordinate member parties' campaigns. But this time there was substantially less money available from the EP budget. Moreover, the EPP Group was much less willing than it had been in 1979 to hand over a part of the available resources for the kind of campaign proposed by the party. Since the winter of 1983–84, there had been an Election Campaign Working Group chaired by the Secretary General, which regularly brought together the responsible workers from member parties to ensure a flow of information and to discuss common projects. The result, as can be imagined, was not a unified election campaign, nor even a coordinated one. Mutual awareness, though, brought very good results in harmonised efforts and specific ideas. Moreover, there were long-term benefits and integration arising from this kind of practical cooperation between experts and senior officials from different national party headquarters.

The results of the 1984 European elections were disappointing. The EPP, far from increasing its lead vis-à-vis the Social Democrats, fell farther behind. After the first direct elections of 1979, the EPP had won 107 seats to the Social Democrats' 112; it was now 110 for the EPP Group, while the Socialists had 130. Once again, however, the EPP had a larger absolute share of the vote than the Social Democrats – 31 million against some

30 million. But the EPP's overall result was two million less than in 1979, and its three-million-vote advantage melted to 750,000.

The 1989 Elections

The 1989 European campaign was prepared and executed in the same way as the campaign for the first direct elections in 1979. Once more a working group met early on, bringing together fairly regularly the responsible member party officials. Given that the institutional context of the election was identical, everything took a familiar course. Nor did the EPP, any more than its rival European parties, have additional resources, let alone new instruments at its disposal. It could do no more than before: put in place the basic political programme and ensure a degree of consensus among the member parties on how to proceed.

The EPP secured 121 seats in this election. This was a relatively low score, given that for the first time parties from Spain and Portugal took part in the election under EPP colours. With modest results in these countries, the Group managed to win only 19 seats, 15 of them for the Spanish People's Party (Partido Popular, PP), which had decided just before the election to withdraw its deputies from the European Democratic Group (EDG) and join the EPP Group instead. The modest success of this election overall was mostly attributable to the ill fortune of the largest EPP member party. Compared with the 1984 election, the German Christian Democrats (Christlich Demokratische Union, CDU) lost nine seats; other EPP parties generally maintained their previous standing. The Social Democrats in Spain and the UK once again had considerable successes. The effect was that the EPP was seriously weakened in comparison to the Social Democrats, with the gap now widened to 121 against 180.[275]

With the accession of new Member States and, especially, with the growing role of the EP in European Community (EC) politics and decision-making, the impact of the elections also increased. More attention to the results of the elections was justified as the Parliament was accorded

---

[275] Documentation of the results of the elections of the EP from 15–18 June 1989 can be found in Werner Weidenfeld and Wolfgang Wessels (eds.), *Jahrbuch der Europäischen Integration 1988/89* (Bonn, 1989), 433ff.

more competences and legislative power. The process started with the Single European Act, but it was the Treaty of Maastricht that, among other things, made the Parliament a co-legislator for the first time, next to the Council of Ministers. With the Treaties of Amsterdam, Nice and Lisbon, the Parliament has become even more important. This applies equally to the groups: their composition, their share of seats and the balance of power among them. The EPP (and from 1999 until 2009 EPP-ED) Group is no exception to that rule.

The 1994 Elections

The results of the 1994 European elections, along with the 1996 by-elections in three new Member States (Austria, Sweden and Finland) confirmed the dominant position of parties organised at European level, above all the Party of European Socialists (PES) and the EPP. This analysis holds even when they or some of their component elements endured substantial losses, as the PES did in France and Spain and the EPP in Italy and the United Kingdom. The three parliamentary groups which could rely on European parties (the PES, the EPP and the European Liberal Democrat and Reform Party, ELDR) together won more than 398 seats; the other seven groups combined could muster only 169.

The EPP's showing in the 1994 European elections not only fulfilled expectations, it exceeded them. The losses in Italy were foreseen: the Italian People's Party (Partito Popolare Italiano, PPI) – successor to the collapsed Christian Democracy (Democrazia Cristiana, DC) – achieved only about 10% of the vote and nine seats. The DC had held 27. In the UK, the Conservative Party was at low ebb and managed to win only 18 seats compared with 32 in the previous legislature. But almost everywhere else EPP member parties were fairly successful. The British and Italian losses were balanced by substantial wins in Germany and Spain. Altogether, full members of the EPP won 125 seats. To that were added the 18 British and 3 Danish Conservative seats, as well as, in Portugal, Fransisco Lucas Pires – who had formerly been in the Democratic and Social Centre (Centro Democratico e Social, CDSp) and was elected on the list of the Social Democratic Party (Partido Social Democratico, PSD) and joined the EPP – and seven French Liberals elected on the common list Union for French Democracy/Rally for the Republic (Union pour la démocratie française/Rassemblement pour la République, UDF/RPR) list led by the Christian Democrat Dominique Baudis.

The EPP grew at the expense of the ELDR with the defection of the French Liberals of the Republican Party (Parti républicain) to the EPP. This trend, started by Valéry Giscard d'Estaing and some of his friends in the previous parliament, continued. But Giscard's manoeuvre was now taking place on a larger scale. He had reached agreement with Jacques Chirac that all Members of the European Parliament (MEPs) who were on the joint UDF/RPR list – the UDF being made up of liberal-inclined members of the Republican Party and the Christian Democrats and centrists from the Democratic and Social Centre (Centre des démocrates sociaux, CDS) who had already joined the EPP – should now join the EPP Group. The EPP Conference of Party and Government Leaders in December 1993 had, on the recommendation of Pierre Méhaignerie, Chair of the CDS, indicated its agreement. However, while the Republicans kept to this agreement, the Gaullists wanted nothing more to do with it after the elections. They broke the agreement by establishing their own group, the Rally of European Democrats (Rassemblement des démocrates européens). The Irish Republican Party (Fianna Fáil) deputies joined this new group, as did the three deputies belonging to the Portuguese CDSp, a party that had been expelled from the EPP in the spring of 1993.

Two further groups of deputies, generally assumed to be ready to join the EPP, decided to stay where they were: the 8 Portuguese PSD representatives who remained loyal to the Liberal Group, and the 26 deputies elected on the Forward Italy (Forza Italia, FI) list. The latter included a number of former DC members who had come together again in the Christian Democratic Centre (Centro Cristiano Democratico, CCD). Before its collapse, the Italian DC had, with the German and Benelux parties, been one of the pillars of the EPP. Prime Minister Silvio Berlusconi's movement had been successful in the European elections, having won over a large part of voters sympathetic to Christian Democracy. He now proposed this movement as the new Italian partner. But for both internal Italian and wider European reasons, this alliance was not made. The Italians established their own single-nation group, Forward Europe (Forza Europa), and awaited developments.[276]

---

[276] See chapter three of the present volume. MEPs from the PSD and FI joined the EPP Group in 1996 and 1998, respectively.

Wilfried Martens, party leader since 1990, had spent the previous years working intensively to deepen the EPP's political programme and had made strenuous efforts to extend the party's influence on political decision making in all the Union institutions, especially in the European Council. Following the European elections he was now elected leader of the parliamentary group. This was the first time – not only in the EPP – that the same person had led both a European party and its parliamentary group. In retrospect, it is significant for the relationship between the party and the Group that it was the party leader who took over as head of the parliamentary group, and not vice versa. For the leader of a parliamentary group to have taken over as party chair would have seemed far more logical, given the great historical and practical political role of the parliamentary groups in the existence and development of European parties. But interestingly, it has yet to happen.

The 1999 Elections

In June 1999 the EPP emerged from the fifth European elections as the largest parliamentary force for the first time since direct universal suffrage. The newly constituted Group was 233 members strong, substantially larger than the PES Group (180 seats). Far behind them were the groups of the ELDR (51) and Greens (48). The Confederal Group of the European United Left/Nordic Green Left had a total of 42 seats, the group of independents in the Europe of Nations Group numbered 30, the Technical Group of independent deputies/mixed group had 18, and the Group for a Europe of Democracies and Diversities, 16. Eight deputies did not belong to any group. The total number of deputies in the EP had risen to 626 following by-elections in 1995, when Austria and the Nordic states joined. Of these deputies, 512 now joined groups representing European parties: the EPP, the PES, the ELDR and the Greens.

Apart from its member parties' electoral victories in their respective countries, the success of the EPP was, above all, the fruit of long-term, systematic cooperation, combined with determined efforts to win like-minded parties as members; this work had until now been carried out in other political frameworks. The reinforcement of the EPP Group with the 12 French deputies of the formerly Gaullist-oriented RPR, the 22 Italian FI deputies and the 9 PSD deputies from Portugal was not only significant in terms of the numerical balance of power. It also marked a strategic breakthrough. The EPP was once again represented in Italy by a large popular party capable of holding the majority, as it had been up until the DC's collapse in the early 1990s. For the first time, the same was true of France and Portugal. The PSD had – until 1996, when it switched to the EPP – belonged to the ELDR. The RPR and FI, by contrast, had never belonged to a European party at all.

The British Conservatives had a great victory in the 1999 European elections. They doubled their number of seats over the 1994 total, and 36 Tory deputies again joined the EPP parliamentary Group. But they demanded an addition to the name of the Group to make clear that it was not solely composed of EPP members, but included others, namely European Democrats, as they wished to call themselves. The name came from a previously existing Conservative Group that had disbanded in March 1992. The change in the Group's name was accepted, but without enthusiasm, since it accorded the European Democrat elements within the overall Group a degree of autonomy of action and behaviour. The name of the parliamentary Group was now the Parliamentary Group of the European People's Party (Christian Democrats) and European Democrats, or EPP-ED Group for short. This change also suited the Italian FI deputies and the French RPR deputies, since neither of their parties had yet been accepted for EPP membership.

Hans-Gert Pöttering succeeded Martens as Group Chair at this time, Martens having decided not to stand for a second term in the European elections. Pöttering's new role recognised not only the talent he had shown in the previous years as a Group Vice-President, but also the powerful position of the German delegation, with its 43 CDU and 10 CSU members; together they made up almost a quarter of the entire Group. Pöttering, supported by the Group's new Secretary General, Klaus Welle, now moved to take advantage of the new strength of the EPP-ED Group to enhance its political profile. That also meant politicising it and polarising debate in Parliament. The first test of this strategy was connected with choosing the new President of the EP.[277]

---

[277] Instead of seeking an alliance with the PES, Pöttering preferred to cut a deal with the ELDR. He proposed the French UDF Deputy Nicole Fontaine as a joint candidate; in return, the EPP-ED promised to support the ELDR Group's candidate in the second half of the EP's legislature.

The 2004 Elections

Also participating in the European elections of 10–13 June 2004 were citizens of the 10 Member States that had joined the EU on 1 May. The EPP once again emerged the clear victor in these elections, winning 268 of the 732 seats (37%) in the new EP. This retained and somewhat increased the EPP-ED Group's relative majority, even though some Italian and French deputies from parties traditionally in the EPP (PPI and UDF) decided to join the Liberal Group, now the Alliance of Liberals and Democrats for Europe (ALDE). The ALDE had 88 seats and was the third largest group, well behind the PES with 202 seats. The Group of Greens/Free European Alliance (42 seats), the Confederal Group of the European United Left/Nordic Green Left (41), the Independence/Democracy Group (36) and the Europe of Nations Group (27) were far behind these front runners; another 28 deputies did not belong to any group.

Not the least of the EPP-ED's strengths after this election was that it was the only group with deputies from every EU Member State. The EPP-ED Group deputies belonged to a total of 44 national or regional parties, of which most were EPP members. The most important exceptions were the British Conservative Party and the Czech Civic Democratic Party (Občanská demokratická strana, ODS), whose deputies made up the ED part of the Group. Pöttering was re-elected leader of the EPP-ED Group.[278] Thanks to its strong position, the EPP-ED Group secured numerous key positions in the new parliament: 7 out of 14 vice-presidents, 9 of the 22 committee chairs and 3 out of 5 quaestors were deputies of the Group.

The 2009 Elections

Despite the fact that the British Conservatives and the Czech ODS had left the Group, the EPP celebrated a clear success in the 2009 EP elections, gaining 265 (36%) out of 736 seats.[279] As a result of severe losses by the PES, the EPP Group became as big as the second and the third largest

---

[278] This time, Pöttering struck a deal with the PES. During the second half of the legislature (2007–09), he became President of the European Parliament himself.

[279] Figures are as of July 2009, before the Lisbon Treaty came into force and the European Parliament counted 751 members.

groups together. The PES Group, renamed Socialists and Democrats for Europe, had 184 MEPs (25%): their losses were compensated for by the Italian Democratic Party (Partito Democratico) joining the Socialists. The third largest group, the ALDE, gained only 84 seats, much less than expected. The British Conservatives, the Czech ODS and some other Eurosceptic parties founded the Group of European Conservatives and Reformists (ECR). They had only one MEP fewer than the Greens/European Free Alliance (55 seats). Next to ECR, another Euro-sceptic and rightwing group was established, called Europe of Freedom and Democracy. They counted 32 members, a few less than the Confederal Group of the European United Left/Nordic Green Left (35). Twenty-seven elected representatives did not ally themselves with a group.

In this election, the EPP won in all the large Member States (despite being in government in a number of them) and performed particularly well in France (29 seats) and Poland (28 seats). It also won half the seats in Hungary, and all MEPs elected on the Italian People of Freedom (Il Popolo della Libertà, PdL) list joined the EPP Group. As in the 2004 elections, the EPP was able to use its victory in the battle over the appointment of the President of the European Commission. With the support of some other groups and MEPs, José Manuel Barroso of Portugal received sufficient votes for a second term. The EPP also secured the presidency of the Parliament. Twenty years after the fall of the Berlin Wall and five years after the EU's 'big bang' enlargement, Jerzy Buzek of Poland was elected as President of the EP, the first from a former Eastern bloc country to hold such a senior position within the institutions of the EU. But the EPP also became stronger inside the Parliament with 5 vice-presidents, 10 committee chairs and 2 quaestors.

Group Chairs

Along with the EPP's President, the Chair of the Group is the most visible and influential figure in the EPP. The Chair is automatically a member of the EPP Presidency and also attends the EPP Summits. So far, the Group has had six Chairs since 1979: Egon Klepsch (1979–81 and 1984–91), Paolo Barbi (1982–84), Leo Tindemans (1992–94), Martens (1994–99), Pöttering (1999–2007) and Joseph Daul (since 2007). They have been linked to the EPP in different ways, each making a distinct contribution to the party's development and the successful efforts to establish a party that functions separate from but close to the Group.

The exceptional commitment of Barbi from Italy was marked by his close cooperation with the EPP. Indeed, Barbi did not differentiate between the two. For him, both the party and the Group were there to serve the end of a single Christian Democratic and European federalist project. As far as he was concerned, it was reasonable to use all available means to bring this about. Barbi was not rewarded for his stance. After his successful work in Brussels and Strasbourg, it emerged that the *correnti* inside the DC had neglected to ensure he had the necessary support for re-election to the EP.

Klepsch enjoyed a longer period in office than any other chair of the EPP Group. He immediately succeeded Alfred Bertrand (1976–77) as leader of the Christian Democratic Group, and in 1979 became the first Chair of the EPP Group in the directly elected EP. After failing to get elected as President of Parliament, he served as Vice-President of Parliament between 1982 and 1984. After that he again took the helm of the EPP Group until he was finally elected President of the EP.

His repeated election as leader over such a long period reflected the prevailing feeling in the EPP Group that he was irreplaceable. Although his way of running the Group was not uncontroversial, he never had to face a rival candidate. His strengths were the incredible degree to which he was always present, both in Parliament and in the Group. He also had a thorough knowledge, down to the technical details, of every procedure or combination of procedures, and the political problems connected with all of them. He relied on the two strongest national delegations in the Group, the Germans and the Italians. The majority of the German delegation was loyal to him, not only out of national allegiance, but also on the basis of friendly relations – relations he was careful to cultivate.

Klepsch's style of leadership in turn dominated the Group's style: he always tried to head off potential controversies that might harm Group unity. He was generally able to avoid conflict by bringing the relevant individuals and groups together, and reaching agreement through establishing a balance of interests. This was how he retained tactical control. At the same time, he pursued the aims of the vast majority of the Group, and he was consistent and clear-eyed about his strategy. Those aims were the federal organisation of the EC, the reinforcement of its democratic and parliamentary components and, finally, the consolidation of the EPP Group's central position. The goal was to ensure that all decision-making in the EP depended on the EPP Group's agreement or participation.

Tindemans enjoyed only a short period in office. He was Group Chair between 1992 and 1994. It was a turbulent time, also in the EP, with the coming into force of the Maastricht Treaty and the first preparations towards enlargement of the EU with countries from Central and Eastern Europe. Tindemans was able to rely on his authority as 'Mister Europe'. He proved notably receptive and sensitive to the particular concerns and problems of the Group. As a former Prime Minister and Minister of Foreign Affairs of Belgium, he was mostly occupied with institutional issues and the foreign affairs of the EU.[280]

After the 1994 elections Tindemans was succeeded by EPP President Martens. It was only the first, and so far the last time that one person simultaneously served as party and Group Chair. Combining two jobs in Brussels and Strasbourg was physically and mentally quite demanding. But Martens successfully exploited the synergies offered by his double responsibility. During his term the EPP Group changed dramatically. MEPs from the PSD and FI joined the EPP Group. Martens also prepared the entry of the French Gaullists. The MEPs from the RPR joined the Group after the 1999 elections. At the same time Martens continued to pursue the pro-integration course of the Group. Praised by his colleagues for the way he managed the Group, he was criticised for his course of action within his own party, the Flemish Christian Democrats (Christelijke Volkspartij, CVP). He left the EP in 1999 but the position of Group Chair was significantly strengthened: for the first time since direct elections, the EPP had the largest Group in the EP.[281]

Pöttering chaired the Group between 1999 and 2007. He belongs to the waning few who have been deputies in the EP from the time of the first direct elections in 1979. His parliamentary work focused particularly on security issues, institutional development and EU enlargement. Pöttering's contributions were always very professional. He has continued his academic work as a jurist and political scientist in parallel, publishing on occasion, usually on European political issues that concern him.

It is a striking sign of Pöttering's spiritual and physical discipline that he remained so active in grassroots politics: as Land Chair of the Europa-

---

[280] See chapter seven of the present volume for his profile as EPP President.
[281] Ibid.

Union in Lower Saxony, as federal Chair of the Europa-Union in Germany, and above all as CDU Chair in his home constituency of Osnabrück. His consensual style of leadership owed much to his roots in these different, interlinked areas, and to the experience he gained in them.

Pöttering's capacity to apply political rigour to problems, his sound judgement and surefootedness, along with his precise knowledge and his gift for words, were essential for leading a large, complex, transnational parliamentary group. In the Group, as in the Parliament as a whole, he won great popularity and respect for his friendly and elegant manner.

Pöttering was succeeded in 2007 by Daul, from France. An MEP since 1999, Daul became known in the Parliament as Chair of the powerful Committee of Agriculture and as Chair of the Committee Chairs' Conference. Unlike his predecessors, Daul was not 'raised' in politics. He developed a career as a farmer and was active within the French agricultural movement. Compared with Pöttering, with whom he shares the same convictions and the same passion for Europe, he is much more a man of practice.[282] Contrary to Pöttering, Daul is able to benefit from the fact that collaboration between the Group and the party has become easier in the aftermath of the British Conservative Party's and the Czech ODS's departure from the Group. This altered situation promises to create new opportunities for the mutual strengthening of the party and the Group.[283]

Other Parliamentary Groups

Deputies or members of EPP member parties have also formed groups in other parliamentary assemblies, which like the EP, form part of the European political system as places of debate, consensus building and consultation. This has happened in the Council of Europe (CoE), in the Organisation for Security and Cooperation in Europe (OSCE), in the Western European Union (WEU), in the North Atlantic Treaty Organisation (NATO) and, more recently, in the Committee of the Regions (CoR). The parliamentary groups in the CoE and the WEU assemblies, both with half a century of tradition behind them, hold far fewer meetings than the EP

---

[282] Pascal Fontaine, *Voyage to the Heart of Europe, 1953–2009*, 507–8.

[283] In his 'Five conditions for future success of the EPP', Fontaine starts with 'Close cooperation between the Group and the Party, each fulfilling its own political function, but serving the same values'. Ibid. 545.

Group does. As a result, there are fewer opportunities for the level of political cooperation that distinguishes the EP.[284]

The Group of the European People's Party/Christian Democrats (EPP/CD) in the Parliamentary Assembly of the Council of Europe (PACE) counts 202 members (out of 636) from 47 countries.[285] The Group is chaired by the Italian Luca Volonte. He succeeded the Flemish Luc Van den Brande in 2010. As with the EPP Group in the EP, there are no representatives from the UK. Since 2008 the EPP Group in the Parliamentary Assembly of the OSCE is chaired by Walburga Habsburg-Douglas from Sweden.[286] The way these parliamentary assemblies and their groups operate obeys the logic of international and intergovernmental cooperation, an approach that does not favour the democratic and federal starting point that characterises the successful construction of a parliamentary group. Nonetheless, the fact that there are groups of deputies from EPP parties is an indication of the vitality of the EPP as a European party.

The EPP Group in the CoR was founded in March 1994, at the same time as the CoR was established under the Maastricht Treaty to create an advisory assembly made up of representatives of regional bodies (Länder, regions, provinces, municipalities and cities).[287] From the beginning, the EPP was the strongest political force in the CoR. Thanks to the federalist and regionalist Christian Democratic traditions, the Group has always been the driving force for the development of this institution, whose essential credo is the subsidiarity principle. The EPP Group in the CoR is made up of regular and substitute members who belong to EPP parties or to parties whose ideas or programmes are politically close to the EPP. Out of 344 regular members, 125 are part of the EPP Group.[288]

During its 15 years of existence, the EPP Group in the CoR was chaired by Jos Chabert of Belgium (while also serving as Minister for the Brussels-Capital region) from 1994 to 1998, Juan José Lucas Jimenez of Spain (President of the Region of Castille-Léon) from 1998 to 2001, Claude de

---

[284] On an ad hoc basis but very intense and successful, EPP representatives in the Convention on the Future of Europe (2002–03) composed an EPP Convention Group, chaired by Elmar Brok. See Wilfried Martens, *Europe: I Struggle, I Overcome* (Brussels, 2009), 169–72.

[285] Figures as of 1 January 2011. See http://www.epp-cd.eu.

[286] See http://www.epp.eu.

[287] 'The EPP Group in the Committee of the Regions', *EPP News* 8 (1995).

[288] Figures as of 1 January 2011. See http://www.epp.cor.europa.eu.

Grandrut of France (Councillor for the Picardy region and Deputy Mayor of Senlis) from 2001 to 2002, Reinhold Bocklet of Germany (Bavarian State Minister for Federal and European Union Affairs) from 2002 to 2003, Isidoro Gottardo of Italy (Councillor for the Friuli-Julian-Venetian region) from 2003 to 2010 and since 2010 Michael Schneider of Germany (State Secretary for Federal Affairs and Envoy of the State of Saxony-Anhalt to the Federal State). Over this period, four of a total of eight CoR Presidents have come from the EPP Group. Each has served two years: Jacques Blanc of France from 1994 to 1996, Chabert from 2000 to 2002, Peter Straub of Germany from 2004 to 2006 and Van den Brande from 2008 to 2010. The EPP Group in the CoR is aided organisationally by a Secretariat led by Heinz-Peter Knapp.

Part III

Party Programme

Chapter Fourteen

Programmatic Profile

From the outset, much of the political activity of the European People's Party (EPP) has been closely linked to its programmatic development. The co-founders of the party, Hans August Lücker and Wilfried Martens, were convinced that solid foundations were needed so that the process of building Europe would not be completely dominated by the preoccupations of day-to-day politics.[289] This idea has lasted, thanks to many inside the EPP who have never ceased to focus on the essentials and on the long term. Since the 1970s, in the wider context of political dealignment or, later on, with the so-called end of ideologies, these efforts have not lost their significance. On the contrary, they continue to play an important part in the development of the party.

Making the party's basic ideas and values explicit has also proven to be important whenever the EPP wants to properly distinguish itself from other political parties and families. This seems particularly true at the European level, since European parties, unlike those at the national level, cannot rely as much as they might want on people – the politicians who are the figureheads of the party, especially during electoral campaigns – or on policies, which in the European context are first of all attributed to the groups in the European Parliament (EP).

---

[289] Wilfried Martens, *Europe: I Struggle, I Overcome* (Brussels, 2009), 36–7. See also, for instance, the opinion of Hanns Seidel in his *Weltanschauung und Politik*, cited in H. Möller, 'Hanns Seidels christliches Menschenbild, H. Zehetmair (Hrsg.)', *Politik aus christlicher Verantwortung*, (Wiesbaden, 2007), 91: 'It is in the parties as well as in the interest of the whole more valuable to have a philosophical guideline rather than the pure technical control of power, free from any philosophy [Weltanschauung], which is then seen as 'statesmanship' [Staatskunst].' (author's translation from German).

The enlargement to include Conservative parties and other like-minded forces made it urgent for the EPP to invest a lot of energy in defining what the political family, currently representing the centre-right in Europe, stands for, especially vis-à-vis the outside world: journalists, competitors and other voices in the political arena. It was also necessary for internal reasons, especially to assist both the new and the old member parties. Having a well thought-out vision of EPP fundamentals allows existing member parties to recognise clearly which political family they belong to and allows candidate parties to understand what they are expected to subscribe to.

Equally important, referring to basic ideas and values when policy proposals have to be drafted – for instance, in the framework of new legislation for the EP – also generates a process of consensus-seeking inside the EPP. Given the diversity of its member parties and the complexity of present-day politics, especially at the level of the European Union (EU), it is, after all, the political programme that unites the party.

The Meaning of Christian Democracy

What specifically distinguishes Christian Democracy from other non-socialist political movements, especially from the Conservatism of the British or Scandinavian type? The question is not just theoretical. It played a prominent role in the discussions surrounding the founding of the EPP, and remained a live issue throughout the party's development because of the recurrent problem of whether or not Conservative parties were to be included or excluded from European Christian Democratic organisations. The question merits attention, since the EPP defines itself as 'Christian Democratic'.[290]

Indeed, if one ignores the way the party is rooted in the Christian Democratic movement, the birth and development of the EPP makes little or no sense. The resistance and doubts that attended the 'strategy of openness', begun in the late 1980s, which brought in Conservative and other parties of the centre-right, make no sense without an understanding of this background. The profound significance of the enlargement process for

---

[290] See 'We, Christian Democrats' in the Basic Programme adopted at the 1992 Athens Congress, and the preamble of the Statutes, in the version approved by the 2009 Bonn Congress.

many Christian Democrats is evident from the fact that even after a basic consensus had been established, the process was not completed for many years.

Here, Christian Democracy is first of all presented as a political movement, that is, a movement deeply rooted in politics. It is also a philosophy, as it states how politics should be conducted and for what purpose, what form society should take and how the state should be organised. Finally, it is also a historically determined identity. Its roots and its basic ideas have travelled a long way through changing political circumstances. Multiple layers of experience have become an intrinsic part of Christian Democracy.

A Political Movement

Christian Democracy is a product of the political history of the countries of continental Europe, where society has largely been shaped by Catholicism and Protestantism. Christian Democracy has remained fairly alien to the political tradition in Britain, and indeed to political thought in both the Anglo-Saxon world and Scandinavia.

In some countries, there have always been national state churches, such as in Britain and Scandinavia. In others, the church has been socially and politically powerful and influential until the most recent times – as in Poland or Spain. In none of these cases has there ever been much need to establish lay political movements bound to the churches and committed to defending ecclesiastical institutions and positions. The same applies, roughly speaking, to the countries of South-Eastern Europe. Here, Orthodox Christianity has created different cultural and social conditions, and the independent national churches have willingly subordinated themselves to those holding power at any given time.

Without exception, the Christian Democratic parties in the Benelux, Germany, France, Italy and a few other countries in Western and Central Europe grew consistently out of confessional movements. These were predominantly Catholic. But some – for instance, in the Netherlands – were Protestant. Such movements, in certain cases already established in the nineteenth century, were reactions against secular or anti-clerical tendencies in the modern state. Later, interdenominational (Christian) parties emerged from these movements. They freed themselves from their earlier defensive attitude, developing their own social agenda in the confrontation with Socialism and Liberalism. In this way, they prepared themselves to take on government responsibility. In this way, too, the reconciliation of the Christian churches and their members with the modern state was made both possible and easier.

On the basis of their experience in government and of the responsibility that goes with power, these parties opened up further. The process accelerated as the century wore on, becoming more pronounced after the Second World War. Forces were integrated into the Christian parties that were not predominantly confessional or ideological (that is, seized of an ethical conviction). Rather, what drove them was a pragmatic or rational impulse (in the sense of ethics of responsibility).[291] Only now could they perceive themselves as Christian Democratic or Christian Social parties, meaning parties made up of citizens inspired by Christianity, citizens who were taking over responsibility in state and society quite independent of their respective churches.

In Britain and Scandinavia, by contrast, the conflict with Liberalism and Socialism produced bourgeois people's parties. These were conservative movements committed to defending the traditional social order, its values and state institutions. In their own countries they represent the same classes of voters, in terms of sociology and value systems, as do Christian Democratic parties in continental Europe. However, they are very different in their party programmes, political approach and style.

A Political Philosophy

At the heart of the Christian Democratic philosophy stands the human being, who is morally responsible before God and has a political responsibility towards society. Christian Democracy is as much concerned with the individual as with human society, as well as with the national and international community. In other words, the focus is not the interests of a particular class or the preoccupations of a specific group. That is why Christian Democratic parties see themselves as 'people's parties', attempting to integrate the differing interests and concerns of all classes and groups by developing consensus. And so this building of consensus, meaning the constant endeavour to strike a balance and reach an understanding between

---

[291] On the ethics of conviction versus the ethics of responsibility, see Martens, *Europe: I Struggle, I Overcome* (Brussels, 2009), 28–9.

divergent efforts and antinomian principles, is also a fundamental element in the Christian Democrat approach.

This is also the driving force behind the concept of the *Soziale Marktwirtschaft* (social market economy), propounded by German scholars and politicians with Christian Democratic leanings. The concept concentrates on squaring economic efficiency with social justice, reconciling the laws of the market with the precept of social responsibility. Similarly, Christian Democrats regard federalism as a suitable method of linking the need for unity with respect for diversity. Thus, they do not envisage the EU as a centralised undertaking. Quite the reverse: Christian Democrats support European integration because they believe that in the present historical circumstances, only integration and cooperation with, and belonging to a broader community will allow the identity of the various peoples and nations to be preserved and to blossom.

The need for integration derives from the fact that almost all social, economic and political problems arising in individual European states today are transnational in nature, and can only be resolved by common endeavour in a spirit of solidarity and in line with the principle of subsidiarity. Subsidiarity and solidarity are two cornerstones of the Christian Democratic credo, which is also strongly influenced by the social teachings of the Christian churches. Catholic doctrine, steadily developed in papal encyclicals on society since 'Rerum Novarum' (1891), is of special importance. Conversely, Catholic doctrine has itself been greatly influenced by Christian Democratic thought and practice. This became especially clear in the papal encyclical 'Centesimus Annus' (1990), which represented the first time that the social market economy, though not mentioned by name, was held up as a concept which confirms the teachings of the church.

The Christian Democratic philosophy has consequences for the political attitude and behaviour of its champions. It calls for openness to dialogue and patient efforts to reconcile opposing positions. Christian Democrats thus find themselves in the middle of the political spectrum, seeking a balance between the paradoxes that permeate the lives of both people and politics. They prefer moderate solutions and a measured approach to any form of radicalism. This attitude involves receptiveness to fresh ideas and flexibility in new situations. It means that Christian Democrats do not follow an ideology even if it claims to have all the answers. They know that no human being is perfect, and that is especially true of Christian Democrats themselves. They do not want to implement some ideological theory worked out by bureaucrats or academics. Their priority is to seek solutions that satisfy people's needs and are therefore accepted by them. So pragmatism, as long as it does not become an end in itself and thus ideological, is a key element in Christian Democratic thinking.

This overview of the Christian Democratic model is in no way intended to idealise the Christian Democratic movement. However, it is important to stress that it is a living model. And it is a model kept in mind by most politically active Christian Democrats with positions of responsibility in state or society. It is certainly not binding on every individual to the same degree, but it does serve as a general guideline for everyone. Evidence for this can be found in the political programmes of the Christian Democratic parties of Europe and of the EPP, and in the political actions based upon this model.

A Historically Determined Identity

Christian Democrats in the various European countries have had different historical experiences, and the same is true for Latin America. It follows that Christian Democratic identity varies too: different individual aspects or elements are valued and regarded differently in different places. Not all Christian Democrats have had the same teachers or read the same books. It is hardly surprising, therefore, that Christian Democrats from different places will acquire varying areas of focus, not only in theory but also in practice. It is evident that the issue of an identity from which a consensus can be forged will be key to any efforts to build up cooperative federations of (or joint efforts by) Christian Democratic parties continent-wide, or even worldwide. What remain the most decisive factors for the shared identity of Europe's Christian Democrats will certainly be common action and the way responsibility is exercised in specific situations. If Christian Democrats continue to regularly invoke the names of Robert Schuman, Alcide De Gasperi and Konrad Adenauer, it is because these personalities have been examples of Christian Democratic identity through their actions and decisions, their successes and achievements. Not one of these 'founding fathers' was a theorist; none of them would have dreamed of defining what Christian Democracy might be. They were Christian Democrats because that was how they saw themselves, and because of how they behaved.

Christian Democracy cannot be understood as an abstraction. It has no 'pure' form. Rather, it exists only as an expression of a movement or as a political programme, and as the action of a party which changes in accordance with the people it represents and the circumstances it confronts. Even the idea or philosophy underpinning the movement has grown out of the conflict with realities within which it has to prove itself. Its historical character can also be seen in the fact that it bears the mark of various (national) imprints and has been through various stages of development. It follows from all this that the ability and readiness to make common cause with political forces of other traditions to achieve common goals is as much of a necessity for Christian Democrats as is preserving their own tradition.

Thus Christian Democracy is by its nature a political philosophy. Historically, it has developed on the back of a political movement, opposing both Socialism and Liberalism. Christian Democratic identity is thus determined not only by its philosophy. The fact that it is a movement and, more particularly, a movement with a history, is just as important. Along with Christian social teachings, both Catholic and Protestant theology and literature and a number of other factors, the successes and failures of Christian Democratic parties and personalities play a crucial role in the search for identity – for the EPP as well.

Drafting the Basics

The importance that the political programme has gained for the EPP is shown by the various occasions on which it was at the top of the party's agenda. A brief look at the list of party Congresses shows the efforts that have been made to regularly draft and update programmatic documents.[292] Roughly speaking, one can distinguish between three sorts of such documents: basic documents that address fundamental ideas and values, irrespective of day-to-day politics (and therefore without a political expiry date, one might say); those that develop certain basic ideas, given a particular political context; and so-called manifestos – electoral programmes that are drafted in the run-up to elections (and are only intended to last or to be valid until the end of the term). In the EPP's terminology, the manifestos are Action Programmes. Since they list the policy preferences and ideas for the forthcoming parliamentary legislature, the EPP Group in the EP plays a

---

[292] See Table 3 in chapter six of the present volume.

crucial role in developing them. Documents with a more specific focus have mainly dealt with institutional reform or, more generally, the process of European integration. Some have also put social and economic policies at centre stage, the latest example being 'The Social Market Economy in a Globalised World', adopted at the Bonn Congress in December 2009.[293]

The party's basic ideas and values have been the primary subject of programme drafting on three occasions. The first time the EPP created a basic document was when it was founded in 1976.[294] The idea of revising the document was launched in the second half of the 1980s. This process came to an end with the adoption of the Basic Programme at the Athens Congress in 1992. In the context of the party's rapprochement with Conservative and other like-minded forces, along with the reunification of Europe, an updating exercise led to the document called 'A Union of Values', adopted at the 2001 Berlin Congress.[295]

Towards a New Basis for Consensus

The process of enhancing cooperation within the framework of the EPP during the 1980s made one thing clear: constant attention to the party's philosophical foundations was of the essence. Without that, the EPP would not over time be in a position to establish the political consensus it needed to act effectively. This was urgent, since these were the years when the EPP was poised to take on ever-greater political responsibilities in the

---

[293] References to the founding principles of the EPP are therefore made explicit, much more when compared to the Action Programmes. See, for instance, p. 3: 'The Social Market Economy is not only a set of economic rules but a deeply normative system based on strong values. The founding fathers of the Social Market Economy envisioned a system in which economic growth and social sustainability were no longer opposed to each other but would work in harmony to ensure a sustainable and positive development for society as a whole. The Social Market Economy was intended to bring the values of individual freedom, solidarity and subsidiarity together. It is a concept that combines the efficiency of the market with equal opportunities, an alternative to classic liberalism, which is based on efficiency without equitable development, and to socialism, which rests on equitable development without efficiency. The Social Market Economy is deeply rooted in our political family. It has found its clearest expression in the development of Christian Democracy in post-war Europe, and it represents a basic principle of all centre-right parties in Europe'.

[294] See chapter two of the present volume.

[295] See the annexes of the present volume.

European Community (EC) as it moved towards a union, as well as responsibilities within various Member States at parliamentary and government levels.

In 1985, Leo Tindemans ordered a joint EPP and European Union of Christian Democrats (EUCD) Working Group to be set up to study 'the spiritual foundations of Christian Democratic politics'. The group worked on the assumption that Christian Democracy had drawn its political force and effectiveness from the situation in Europe after the Second World War. People who had been liberated from Fascism and spared from the oppression of Communism found that the Christian Democratic value system suited their social, political, economic and cultural situation. But there had been, since then, a change of mood because of a constant shift in values.[296] Following the youth and student revolts of the 1960s, the diminished influence of the church, widespread questioning of the validity of the Western political order and numerous other crises had all contributed to the erosion of almost every tradition. Christian Democracy had lost much of its original magnetism.

Another factor was that the inevitable pressures of gaining and holding power pushed basic questions about the nature and direction of policy into the background. That was also true for the Christian Democratic parties, which had enjoyed extraordinary success in the 1950s and 1960s. The result was a loss of self-confidence. In constant fear of being pigeonholed as intolerant, intransigent or reactionary, the Christian Democratic movement was reticent about mounting a proper defence of its values. Such a retreat was especially marked during the 1970s, when Social Democratic doctrine became more and more dominant.

Things were not improved by a reluctance to consider theoretical questions and to formulate a specifically Christian Democratic 'doctrine'. Such weakness in terms of political theory is no doubt the flipside of the strength of Christian Democracy in Europe: even in the heyday of ideology, the movement had never fallen into the ideology trap. Fortunately, the EPP was at the time chaired by a politician who in his personal development and his career had always recognised the importance of the founding principles of Christian Democracy. It was therefore no coincidence that

---

[296] See in this context the works inspired by the European Values Study Group (EVSG), e.g. Sheena Ashford and Noel Timms, *What Europe Thinks: A Study of Western European Values* (Aldershot, 1992).

Tindemans had launched the process of re-examining Christian Democratic foundations.

The Spiritual Foundations Working Group, chaired by EPP and EUCD joint Secretary General Thomas Jansen, resolved at its inaugural meeting to mount a series of colloquia on the problems of Christian Democracy in detail and depth. The first took place in Amsterdam in April 1986. As the basis for subsequent work, the initial task was to examine the tradition of Christian Democratic thought. Questions such as the limits of the welfare state and the moral and practical aspects of development aid would be covered prior to considering Christian Democracy in practice.[297] Further colloquia followed: on 'International Solidarity' (held in Praglia, northern Italy, in autumn 1986); 'Federalism as an Element of Christian Democratic Thought' (Vienna, spring 1987); 'The Social Market Economy: A Christian Democratic Idea' (Paris, autumn 1987); 'What Is a Christian View of Humanity', and 'What Political Relevance Does It Have?' (Madrid, summer 1990), and 'The Future of Europe' and 'The Contribution of Christian Social Teaching' (Chantilly, autumn 1990).[298]

In 1989, however, the political upheavals in Europe and the world in general meant rethinking and reformulating positions that had been held by the EPP since its founding. This was a practical political necessity. The international context of the European unification process had altered with the ending of the ideological conflict between East and West. That had implications for political priorities. At the same time, the climate of opinion had altered. The change enshrined in the Maastricht Treaty, the evolution from the EC to the EU, also meant a reappraisal of ethical values and a political redefinition.

Still, the EPP did not want to jeopardise its Christian Democratic identity or its role as a people's party occupying the political centre. That intention applied to the vast changes taking place from 1989 onwards,

---

[297] See Piet Bukman et al., *Tradition und Aktualität der Bemühungen um eine 'Doktrin'*. Vol. 1, EPP series Geistige und historische Grundlagen Christlich-demokratischer Politik (Melle, 1988).

[298] See also Thomas Jansen, 'Europäische Christdemokraten überprüfen ihre Doktrin', *Politische Meinung* 256/36 (March 1991), 66–72. These seminars in conjunction with an allied institute or foundation were prepared and put into effect to create a network.

and especially to the issue of opening the party's doors to other, notably Conservative, political forces. A Christian Democratic identity was the wellspring of the party. And it was assumed that the EPP's position at the political centre would also be the main attraction to the Conservatives. The EPP then was prepared to join forces with the Conserva-tives only on the condition that the party's essential Christian Democratic direction – its Christian image of man, the *Soziale Marktwirtschaft* and in particular its federalist European stance – were not brought into question.

The 1992 Athens Basic Programme

The EPP party and government leaders had called for a new basic document to be drawn up. In April 1991, they decided that while the party would be opened to the British and Scandinavian Conservative parties, the EPP's Christian Democratic identity should also be deepened. The greater the range of parties working together, from different traditions and ways of thinking, the more urgent it would be to accentuate the party's identity to ensure political cohesion and effective common action.

The resolution of EPP party and government leaders on 13 April 1991 affirmed that the EPP saw itself as 'a Christian Democrat-inspired force of the political centre' that would 'make determined efforts' to preserve and develop its Christian Democratic identity and political programme.[299] The Political Assembly duly decided on 3 July 1991 to set up a Basic Programme Committee with the following mandate: 'The Basic Programme Committee will elaborate a new Basic Programme reflecting the consensus reached in the EPP. The Basic Programme will place in practical context the programmes agreed at EPP Congresses, and develop them further . . . The Basic Programme Committee will concentrate on articulating and formulating the principles which guide the EPP in specific [policy] areas.'[300]

---

[299] Procès-verbaux, EPP Conference of Party and Government Leaders, Brussels, 13 April 1991, EPP General Secretariat Archive.

[300] The mandate also lists the texts which the Commission should strive to put into action and develop: Political Programme 'Together towards a Europe for Free People', 1978; Manifesto of the European Christian Democrats (EUCD), 1976; Political Manifesto of the Christian Democratic World Union (CDWU), 1976; the Action Programme for the Second Parliamentary Term of the EP (1984-1989), 1984; 'On the People's Side', Action Programme for the Third Parliamentary Term of the EP (1989-1994), 1988; and 'For a Federal Constitu-

This work was so important it was decided that the Basic Programme Committee Chair should be EPP President Martens. The fruits of the Committee's labours would be presented to the autumn 1992 Congress to be debated and voted on. The Committee, which included representatives from member parties, recognised associations and the EPP Group in the EP, met several times between autumn 1991 and autumn 1992. Its discussions were based on texts for individual chapters or paragraphs that had been proposed by Jos Van Gennip, Kees Klop, Paul Maertens, Thomas Gauly, Paul Dabin and Léon Saur.

Originally it had been thought that policy areas of key importance to European integration should, along with other matters, be dealt with under the heading 'Areas of application'. These included foreign and security policy, development aid, social and economic policy, and policy towards business, including domestic, environmental and cultural policy. However, this idea was abandoned during the course of discussions. The Basic Programme, it was decided, should not be cluttered up with specific policies. Rather, these should be dealt with later in an Action Programme.

The first chapter of the Basic Programme contained a new, up-to-date description of Christian Democratic philosophy, based on the Christian view of man. It set out the philosophy's basic values and what they meant in terms of ideas about society and the political system. A fresh description of the basics of Christian Democratic thinking had become necessary in light of the new socio-political order in Central and Eastern Europe after the watershed of 1989, and the domino effect it was having on European integration policy. But another reason was the change in Western European values since 1968. By setting out the principles that were supposed to inform EPP policy, a picture emerged of the Christian Democratic conception of the way in which society and the political system as a whole should evolve.

The second chapter was devoted to the EU, laying out proposals for its future federal and democratic constitution. This chapter contained the EPP's federal European creed and its vision of the future shape of the EU. At the time of the Congress the Maastricht Treaty had been negotiated and

---

tion for the European Union', 1990. Published in the preface of the documentation for the Basic Programme Committee, 'In preparation for the IX EPP Congress 1992: basis of the programme', EPP General Secretariat.

signed, but not yet ratified. The referenda in Denmark and in France had shown how extremely difficult the ratification process would be. However, the Congress identified the Maastricht Treaty as the starting point for all further developments.

These two chapters were introduced by another chapter penned by Martens, in which he analysed the situation of European society and the challenges it posed. From this starting point, he attacked the various – new and old – ideological temptations. It was a way of pointing out, at the same time, emerging dangers such as population growth, the increasing interdependence in the world economy, the gap between rich and poor, the exhaustion of natural resources, the revolution in media technology, the growth of criminality, and the resurgence of racism and nationalism. Against all these dangerous developments Martens set out

the chances of a lasting peace after the end of the east-west conflict, and the possibilities of the consensual western model which connected the Social Market Economy with democracy . . . Christian Democracy wants to contribute to this development. Firmly rooted in its traditions, the movement wants 'the best' and seeks to evoke man's constructive qualities by giving up-to-date expression to our ideals of Christian-social personalism.[301]

The consensus recorded in the new Basic Programme was to a large extent the achievement of Martens, whose authority and commitment were applied as a mediator, as well as in his decisive role in the Programme Committee and the Congress. His contributions were essential to the agreements reached and to the quality of the results. In Martens's own words:

Renewing the Basic Programme was for me a matter of fundamental importance and a high priority. I was convinced that the expansion of the party would only prove durable and fruitful if there was agreement about the party's political foundations. Moreover, the acceptance of these fundamental principles had to be a basic condition for membership for new political parties. It was a fact that the greater the difference in parties the more important the common basis became. This was our only true *raison d'être*, for one does not succeed through *Realpolitik* alone.[302]

---

[301] Published in *Europa 2000: Unity in Diversity*. Vol. 9, EPP Documentation series, (Melle, 1994).

[302] Martens, *Europe: I Struggle, I Overcome* (Brussels, 2009), 123.

Dissent

Even after the adoption of the Basic Programme, many of the old Christian Democratic–oriented members were concerned that the Conservative and pragmatic representatives of the new member parties might be trying to water down the party's fundamental positions. One sign of this was the establishment of an informal 'Athens Group', made up of political figures from several countries. Its object was to defend the Athens Basic Programme and traditional Christian Democratic policies in general against actual or supposed attempts to reinterpret them. The Athens Group was established in the aftermath of the entry into the EPP Group of the Members of the European Parliament (MEPs) of Forward Italy (Forza Italia, FI). It made its voice heard from time to time, notably in the EPP Group, but also in party bodies. Its activities, however, did not last long.[303]

Despite the best efforts of EPP party and parliamentary leaders, the divisive arguments connected with finding a new consensus were to rage on for quite some time. One of the most spectacular episodes in this long-running conflict was the deep crisis of confidence which hit the EPP in spring 2000, when EU heads of state and government condemned Austria over the coalition government between Wolfgang Schüssel's Austrian People's Party (Österreichische Volkspartei, ÖVP) and the extreme-right Austrian Freedom Party (Freiheitliche Partei Österreichs, FPÖ). After the problem had been solved, resentment still existed among some and led to the establishment of the 'Schuman Group', another attempt to uphold the basic principles of Christian Democracy. This group operated mainly from within the EPP-ED Group, but none of its initiatives came to fruition.[304]

For some members of the Schuman Group the story was to come to an end. Two EPP founding member parties, the Italian People's Party (Partito Popolare Italiano, PPI), formerly Christian Democracy (Democrazia Cristiana, DC) and the Union for French Democracy (Union pour le démocratie française, UDF), formerly the Democratic and Social Centre (Centre des démocrates sociaux, CDS), decided to withdraw their deputies from the EPP-ED Group following the 2004 European elections. This in essence meant leaving the EPP. The pervasive sense of crisis was qualified, if only somewhat, by the fact that these manoeuvres were above all connected with domestic politics in France and Italy: the ideological issues were

---

[303] Ibid. 147.
[304] Ibid. 164–6.

essentially a front. The PPI, in opposition in Italy against Silvio Berlusconi's government, no longer wished to work with his party, FI, at European level. As for François Bayrou, the UDF leader, who was said to entertain hopes of becoming president of France, he did not – given this ambition – want to be chained to the Union for a Popular Movement (Union pour un mouvement populaire, UMP), the governing party. Yet this episode can be dismissed perhaps as an endgame. Intensive work on policy had in the meantime produced a substantial sense of party togetherness.

'A Union of Values'

In the meantime, the EPP had started to update the Basic Programme without, however, wanting to replace it. The outcome was to be considered an addition, rather than a substitute. A draft document, which served as the basis of discussion in the four Working Groups and in the plenary session of the Berlin Congress, had been agreed on in the Programme Committee chaired by Martens. The Committee's task was to elaborate a paper for ideological renewal. This was intended to update the political programme in light of the continuing debate about the future European Constitution and the Union's imminent 'big bang' enlargement. The backdrop to all this was the continuing effort to achieve integration and consensus between the old and new member parties, each with its particular history and character.

The comprehensive document agreed upon at the Berlin Congress is set out in six chapters: 'People at the Centre of the Union', 'Europe's New Economy in the World', 'Challenges of the Information Age', 'European Identity in the 21st Century', 'A Europe Open to the World' and 'New Approaches and Firm Values'.[305] Compared with the Athens Basic Programme, this document contains little emphasis on philosophical or theoretical roots, or on describing the underlying rationale for the principles guiding the politics of the EPP.

At the Berlin Congress, the EPP was much more assertive than it had been in previous documents. The evidence is especially clear in the reference to guiding principles and values for political action. But the concept of values is very broadly interpreted, embracing a wide range of goals – material, economic and social. These are posited as relevant to creating the right conditions for the fulfilment of the individual and of European so-

---

[305] 'A Union of Values', adopted by the EPP Berlin Congress, 11–13 January 2001. See the second annex of the present volume.

ciety as a whole. In the document the EPP sets out a total of 617 points describing what it intends to achieve, what the party is for and what it is against. In that sense this 'resolution' (adopted by the Congress) comes close to an action programme.

Unity in Diversity

Work on a political programme, being a matter of forging a consensus, also tends towards political integration. Common programmes contribute to European integration and to uniting peoples and states. The same is true of the efforts needed to formulate those programmes, in which overcoming national or cultural differences is an inescapable element. The need for integration and agreement does not apply only to policies. It also applies to organisational issues and to the party's statutes, as well as to its basic ideas and values. And on such matters the various components of the EPP often have their own – differing – opinions as well.

Efforts to nonetheless find agreement and develop a clearly defined party programme have been crucial to the EPP's development and have involved all party bodies. By focusing on its political programme, the party creates unity in diversity.[306] Establishing consensus between the member parties and their leaders is central to all EPP undertakings. That consensus will initially be forged between member parties' leading figures and elected representatives, but eventually between their active members as well. Indeed, such consensus is part of why the EPP was founded in the first place. A European party can ultimately achieve nothing if it does not know – or cannot articulate – what it seeks to achieve.

---

[306] 'Unity in diversity' (in Latin *In varietate concordia*) is the unofficial motto of the EU.

Political Programme, Adopted by the First EPP Congress in Brussels on 6–7 March 1978

European People's Party

FEDERATION OF CHRISTIAN DEMOCRATIC PARTIES OF THE EUROPEAN COMMUNITY

POLITICAL PROGRAMME

Our guidelines for Europe

1. Our concept of man
2. Human rights and basic freedoms
3. Solidarity and justice
4. Political democracy
5. Culture—the basis of our European identity

Europe in the world

6. Responsibility and solidarity
7. Our alliances
8. Relations with the countries of Central and Eastern Europe
9. Europe and the Third World
10. Europe and the United Nations

The European Community's policy

11. Towards a liberal and socially just policy

---

[307] The EPP political programmes are re-printed here in their original form. They may contain grammatical and stylistic errors and inconsistencies.

12. Economic and monetary policy
13. Social policy
14. Structural and regional policy
15. Transport policy
16. Agricultural and fisheries policy
17. Trade and industry
18. Energy policy
19. Environmental protection
20. Consumer protection

The Community's institutional framework
21. European union
22. The Community organs
23. Advisory bodies

Our goal: a united Europe

OUR GUIDELINES FOR EUROPE

The aim of the EUROPEAN PEOPLE'S PARTY (Federation of Christian-Democratic Parties of the European Community) is European unity. This represents an extension of the successful policy adopted by the Christian-Democratic statesmen, Robert SCHUMAN, Alcide DE GASPERI and Konrad ADENAUER, who laid the foundations for what has already been achieved. We are accordingly firmly resolved, by the creation of a European union, to continue and complete this historic undertaking, the political aim of which is a Federation of Europe, as proposed by Robert SCHUMAN on 9 May 1950.

We have put an end to the conflicts between our peoples which led to the Second World War. Guided by our Christian ideal, as exemplified in the 'European Manifesto' and 'World Manifesto', we aim to continue along the same road and to give the unity of Europe a form which will eliminate the threat posed by the totalitarian powers and allow Europe to fulfil, at world level, its joint responsibility for the maintenance of human dignity.

Only by joint action can Europe safeguard its own personality (its identity), its right of self-determination, and hence its ideals of freedom, solidarity, justice, peace and democracy.

The European union must be open to all European states which recognize these principles and the political objective of unity. In a federation, Europe will achieve this unity and safeguard its diversity.

The unification of Europe and the achievement of a free, democratic and socially just community is a challenge for the citizens of Europe, and in particular for European youth. It is very important to let young people participate actively in the construction of Europe, for it is their future. The first direct elections to the European Parliament represent a decisive step towards the achievement of these objectives.

The new Europe will have its own personality. It will be:

- *dynamic* in realizing its Christian ideals and belief in human endeavour,
- *strong* by virtue of its freedom, justice and democracy, and
- *constructive* by virtue of its creative power, its international solidarity and its desire for peace.

This European union must be a community in which all the forces of democracy find freedom of expression and in the formation of which such forces can actively participate. Thus, no political movement should be allowed to claim Europe for itself alone in order to enforce its own social system. Any move in this direction would jeopardize the freedom of a truly pluralist Europe.

Europe must remain open to the world and make its own essential contribution to the fight against hunger, poverty and violence and to the achievement of justice and true peace in freedom.

1. Our concept of man

1.1   Our policy is based on a concept of man which is characterized by the fundamental Christian values and finds its expression in the inalienable and inviolable dignity, freedom and responsibility of man—in equality in diversity, in the struggle for self-realization and in the awareness of the fallibility of man.

1.2   Man is dependent on the community for his development and cannot therefore attain to self-realization unless he bears responsibility for himself and for others. The cornerstones of our society— freedom, justice, solidarity, pluralism and openness—all have their place in this personal concept of man.

1.3   In accordance with this philosophy, we confirm the value of the family, which is the mainstay of our society, which is particularly suitable as a means of furthering the development of each of its members and which is a crucial element in the education of children.

2. Human rights and basic freedoms

2.1   We will safeguard human rights and basic freedoms as a foundation for the development of the individual and for the establishment of a just society. We believe that these rights and freedoms must be respected throughout the world. They are set forth in the European Convention on Human Rights, they are laid down by the national constitutions of our Member States and they are formally confirmed by the European Community:

- the right to life, to protection from inhuman or degrading treatment and from unlawful arrest
- the right to liberty and security
- the right to freedom of thought, conscience and religion
- the right to freedom of expression
- the right to freedom of peaceful assembly and association
- the right to take part in free and secret elections
- the right to respect for one's private and family life, home and correspondence
- the right to education and the right of parents to choose the kind of education to be given their children
- the right to own property
- the right to protection against discrimination
- the right to freedom of movement
- the right to leave any country, including one's own
- the right to enter one's own country and immunity from expulsion therefrom
- the right to a hearing by an independent and impartial tribunal
- the right to legal remedy in the event of violation of the fundamental rights and freedoms.

2.2   We also advocate the creation of conditions which will make it possible for everyone in our modern society to benefit fully from these human rights and basic freedoms, and in particular the right to a minimum income, to treatment in the event of illness, to work, to strike, to equal pay for equal work, to a healthy environment, to accommodation, to access to freely chosen educational establishments and the right to objective information.

3. Solidarity and justice

3.1   The solidarity which we are striving to achieve represents a bond between all men and women. It is a prerequisite for community life. Solidarity finds expression in rights and duties. Everyone has a right to participate in the community; everyone has a duty to do everything in his power to help to ensure that the community as a whole can accept responsibility for the individual.

3.2   Our social policy is based on the principles of solidarity and subsidiarity. This means helping others to help themselves and implies a duty to ensure that social justice is maintained. The political community must give individuals and organizations scope to develop to the full their capacity to operate on their own responsibility.

3.3   In this spirit we are ready to fight against injustice, discrimination and poverty. The social problems of inequality between social groups, regions and countries can be solved only by a European policy of solidarity and by structural change.

3.4   We have a particular duty to the socially underprivileged. Our aim is also to give such people the opportunity to develop their personalities and to participate in our prosperity and common progress.

4. Political democracy

4.1   Pluralist democracy is the form of government which best corresponds to our concept of a modern society based on partnership. It finds expression in controlled power over a limited period and backed up by a true constitution. It makes it possible, in accordance with the wishes of free citizens, to introduce adjustments and changes without force and to eliminate tension in society through the self-determination of the individual in the interests of the community as a whole. This calls for the respect of the rights of social, ethnic, ideological and religious groups.

4.2.1   The political parties bear a measure of special responsibility for the promotion of representative democracy and for the greater involvement of citizens in the decision-making process. To enable them to discharge their responsibilities to the community in full, their constitutional status must be legally recognized, clearly defined and safeguarded by the allocation of public funds. We support electoral systems which take due account of the pluralism of political movements but at the same time favour the formation of larger parties in order to guarantee stable and strong government. Since society in its modern form tends to encourage increased governmental power, parliamentary rights of supervision must be strengthened.

4.2.2   Our aim is to strengthen democratic institutions at regional and local authority level and to promote the dialogue between citizens and individuals with political responsibility. We will increase the participation of social groups. At political level, however, their decisions cannot act as a substitute for the decisions of the democratic institutions.

4.3   We also believe that political power should be decentralized wherever possible. In accordance with the principle of subsidiarity, States, regions and local authorities should be empowered to decide on, carry out and supervise everything which can best be done at their own particular levels. In particular, local government must be strengthened. It offers citizens the best opportunity for influencing living conditions in their immediate neighbourhood.

5. Culture—the basis of our European identity

5.1   One of the principal aims of culture is to interpret the world creatively and critically; it should help men, as individuals and in their community, to understand and master their destiny. The wealth of European culture, with its diversity and unity, forms and basis for cooperation between the peoples of Europe in a united Europe.

5.2   We will preserve and extend our cultural heritage, and give contemporary art more scope for development and bring it within reach of all sections of the population. In particular, we will make national and regional cultures more accessible to each other and in this way further European culture as a living reflection of its Christian and humanist traditions.

5.3   We see the stimulation and full development of culture as a golden opportunity for counteracting the tendency towards uniformity and lack of commitment in society, growing materialism and the spread of Marxist ideology. General facilities for extracurricular and further education are an essential means to this end.

5.4   We see the representatives of the arts and the sciences as important partners in the construction of a humane society. The competent public authorities should ensure that our historical, cultural and artistic heritage is preserved and protected. In the process they should support private organizations.

5.5   The curricula of the various types of schools and universities should gradually be harmonized to that diplomas can be recognized in all Member States of the Community.

5.6   Schooling and further education, in particular in history and languages, in the arts and the sciences, should be used to promote European awareness and thus to encourage the emergence of truly European citizens.

EUROPE IN THE WORLD

Only if Europe is united will it be strong enough to further effectively the cause of freedom, solidarity, peace and justice for the peoples of the world.

Only if Europe is united will it be able to make a real contribution, as a motive force behind progress and international social justice, to the creation of a new world order.

Only if Europe is united will it be strong enough to fulfil its responsibilities, to look after its legitimate interests in the world, to assert itself against the threat of military action and to safeguard its future existence as a free and sovereign power.

Now more than ever, at a time when the process of unification is not complete, Europe must 'speak with one voice'. European Political Cooperation which has developed in the spirit of the Treaty of Rome is an encouraging step in this direction. We must build on this foundation. We therefore support the TINDEMANS proposal for a common foreign policy and for the establishment of a decision-making centre to deal with matters of foreign policy.

6. Responsibility and solidarity

6.1   Europe should not be satisfied with strictly safeguarding its own particular interests. Its historical role, its solidarity, its cultural and spiritual calling and its economic power invoke its responsibility.

6.1.1   We have a duty to defend human rights, basic freedoms and the rights of peoples. Unless these rights and freedoms are respected, true peace is impossible. Human rights and basic freedoms are regarded throughout the world as more important than the right of sovereignty. The protection of such rights and freedoms cannot therefore be interpreted as interference in the sovereign rights of another country, particularly when governments have formally committed themselves to respecting them. In this spirit we shall fight against tyranny.

6.1.2   We have a moral and human responsibility towards the countries of Eastern Europe with which we have common historical bonds. We will overcome the division of Europe by peaceful means. The right to self-determination of all European peoples, including the German people, remains for us a principle of European policy.

6.1.3   Europe must also assume responsibility for the Third World. We see cooperation in achieving a better balance of interests between the industrialized and the developing countries, in other words the achievement of a more just world economic order, and the provision of more funds for development aid as priority aims in a European policy of solidarity.

6.1.4   Finally, Europe has a responsibility to help safeguard world peace. We will therefore contribute to the peaceful solution of conflicts. However, we must also make it clear that we are prepared to defend ourselves and protect our independence.

7. Our alliances

7.1   *Our partners in Europe and the Mediterranean*

7.1.1   We confirm the Rome Treaty's undertaking to keep the Community open to all democratic countries in Europe. We support the Greek,

Portugese and Spanish applications for membership of the Community. Culturally and historically these nations belong to Europe. We advocate negotiations aimed at providing these three countries, as soon as possible, with the opportunity to participate politically and on an equal footing in all the Institutions of the Community, even though the economic problems which undeniably exist can only be solved by means of transitional arrangements. While negotiating the treaties of accession we must at the same time take steps to safeguard the further development of the Community and the consolidation of its Institutions.

We advocate close cooperation with the other democratic countries of Europe. Openness and fairness in our dealings with all such partners, the evolution of common positions, and solidarity in times of crisis are our answer to the dangers facing free Europe as a whole. The Community and its members must continue to cooperate fully within the Council of Europe—our link with non-member states.

7.1.2 The Mediterranean countries are also our partners—by virtue of agreements which must be further extended. Peace in the Eastern Mediterranean is a matter of vital importance. In the context of political cooperation, Europe must make its contribution to the search for a balanced solution to the problems of the Middle East which takes account of the legitimate rights and interests of all countries and peoples in that region.

7.2 *The United States of America*
We want the European Community and, later, the European Union to participate as an equal partner in a constructive dialogue with the United States. Relations between Europe and the United States are characterized by broad agreement on the value of freedom and justice, by considerable similarity in political objectives and by the Atlantic alliance, which is essential for the security of the entire Western world.

7.3 *The democracies outside Europe*
7.3.1 In its relations with other Western countries, Europe must strive to achieve solidarity and common standpoints, especially in times of crisis.

7.3.2   Together with the other democratic states it must commit itself to solving the major economic problems of the world, strengthening the pluralistic democracies in their confrontation with totalitarian regimes and protecting the fundamental human rights and freedoms.

8. Relations with the countries of Central and Eastern Europe

8.1   *Balanced détente*
8.1.1   We firmly advocate effective and lasting detente on a balanced reciprocal basis. The furtherance of human, political, economic and cultural relations between the peoples of East and West contributes to the credibility of détente and is thus an important factor in the maintenance of peace.

8.1.2   Berlin (West) is part of the European Community and continues to be one of the barometers of real détente in Europe.

8.2   *The Conference on Security and Cooperation in Europe (CSCE) and its results*
The Final Act of Helsinki represents an important step towards détente between East and West. All the provisions of this Final Act must be fully respected and, where possible, extended. The implementation of the provisions of the Final Act on human rights will make an important contribution to the 'humanization' of the living conditions of the peoples of Eastern Europe. We fully realize that these people are relying on our solidarity.

8.3   *Greater security*
8.3.1   Considerable importance attaches to mutual balanced and controlled troop and arms limitation, particularly in view of the growing military potential of Eastern Europe.

9. Europe and the Third World

9.1   We will cooperate on the formulation of a comprehensive new development and growth strategy in which the developing countries will participate on a basis of equality; Europe can contribute its economic and social experience to this dialogue. To this end we must make the peoples of Europe aware of the real importance of the problem.

9.1.1   The developing countries must use this opportunity responsibly in order to encourage the evolution of a better, and more just system in their parts of the world.

9.2   *The Community's policy based on the Lomé Convention*

9.2.1   Worldwide development policy does not exclude the possibility of special arrangements for certain regions. In the Convention of Lomé, the European Community and about 50 associated developing countries in Africa, the Caribbean and the Pacific embarked on a new form of cooperation based on partnership and equality which is already bearing fruit and which, in a sense, sets an example for cooperation throughout the world between industrialized and developing countries.

9.2.2   The ideals embodied in the Lomé Convention are an essential prerequisite for the creation of a climate of confidence, without which cooperation cannot truly succeed at world level.

9.2.3   In this connection, we believe it important for special relations to be established, in particular with the peoples of Latin America, with whom we have many intellectual and cultural ties.

9.3   *Expenditure, trade and international division of labour*

9.3.1   The industrialized countries—including the Community—must contribute to public development aid at the agreed level (at present at least 0.7% of the gross national product) within prescribed time limits. Steps must be taken to ensure that the funds available are allocated primarily to the poorest nations. It should not, however, be forgotten that the development aid provided by the private sectors in the industrialized countries of the West makes a particularly effective contribution. Favourable conditions for the constant growth of private development aid should be maintained and extended.

9.3.2   Acceptable solutions based on individual cases must be found without delay for the pressing problem of increasing debts, in particular those of the poorest developing countries.

9.3.3   Multinational undertakings must contribute effectively to national development programmes in the developing countries in order to help satisfy the vital needs of the populations of the countries concerned and to encourage socially acceptable developments.

9.3.4   As regards trade in raw materials, international cooperation policy must pursue the following aims, as far as possible within the framework of market solutions: stabilization of the revenue of developing countries by means of compensatory mechanisms, the improvement of market structures and the diversification of production and economic structures.

9.3.5   We recommend that, in the relevant conferences, the debate on the financing of raw materials supplies should be continued and that steps should be taken to ensure that results are obtained which favour the necessary stabilization of particularly important markets.

9.3.6   In accordance with the Community's development policy, the division of labour between industrial and developing countries must be improved. This involves enlarging the scope of the processing industries, extending access to the European markets to products from the developing countries and extending generalized preferences to further products in the context of the Community's commercial policy.

9.4   *Priority for food problems and the promotion of agriculture*
9.4.1   In the short term, the world's food supplies must continue to be secured by making reserves available and by implementing food programmes by means of cooperation in the appropriate international organizations, between countries in surplus and countries in deficit.

9.4.2   Increased food production must be stimulated by the promotion of agriculture in the developing countries themselves, an aim which must also find expression in aid programmes, projects and technical and financial assistance.

9.5 *Greater balance between bilateral and Community development aid*
We believe that the Community should take over more responsibility for bilateral aid provided by the Member States. In this connection funds must be made available to the Community which are more in keeping with its responsibilities. The development aid provided by the Member States must be increasingly coordinated by the Community's institutions.

9.6 *Non-governmental organizations*

Private initiative plays an important role in this field. We draw particular attention to the unique human value of the work of recognized non-governmental social, cultural and religious organizations involved in development aid. The Community must step up its financial support to these organizations and remove administrative difficulties which hamper them in their work.

10. Europe and the United Nations (UN)

10.1 If Europe wishes to be a credible ally of the young nations, it must help to strengthen the United Nations. In close cooperation with the Western democracies and its other partners, the Community must advocate more effective decision-making procedures, stricter and less wasteful running of the secretariats of international organizations, and material decisions which are just as much in keeping with the interests of the young nations as they are with its own ideals.

10.2 Certain new issues transcend the sovereignty of individual states, for example, the setting up of a Council for the defence of human rights, measures against international arms trading, measures for the protection of nature and the environment, measures to prevent the pollution of the seas, measures on the use of the seabed and of space, and the drafting of a code of good conduct for multinational undertakings. Such issues call for a coordinated international policy and the gradual establishment of a recognized international legal system. The Community must contribute to the achievement of these objectives on the basis of its own ideals, legal principles and vital interests.

THE EUROPEAN COMMUNITY'S POLICY

11. Towards a liberal and socially just policy

11.1   Christian-Democratic policy has made a decisive contribution to economic and social progress in Europe. We take as our basis the fact that, for the time being, the economic and social integration of Europe will be a matter for priority. However, we also firmly believe that such integration must be accompanied by corresponding political progress. Economic development is not an end in itself. It centres on mankind. Its aim must be to improve living conditions for everyone and protect the quality of man's natural and cultural environment.

11.2   As a basic prerequisite, the efficiency of the social market economy must be maintained. This is a dynamic economic and social system, the efficiency of which depends on the principles of responsible freedom, the initiative of the individual, the creative power of everyone working in the economy, and competition on the free market. In this system social responsibility finds its expression in active solidarity. It transcends capitalism and collectivism. It ensures a maximum of co-determination and co-responsibility, the development of personality, property and prosperity, and social security for everyone. It is a system in which decision-making powers as well as property are better distributed. It aims to provide full employment and is the most suitable instrument for harmonizing growth with the maintenance of a healthy environment and the sparing use of limited supplies of raw materials.

11.3   The main aim is to create and safeguard employment. In this connection working conditions should be brought into line with technical, economic and social developments. Work is essential to the self-realization of man. Full employment cannot be divorced from the need for effective measures to counter inflation and for satisfactory economic growth.

11.4   In our view a society of this kind is a better way of gradually eliminating unwarranted economic and social inequalities between countries, regions and individuals. This economic system is based on a variety of independent decision-making centres and permits a better distribution of resources. Moreover, effective national action against cartel formation and excessive concentration is essential. Particular attention should be paid to small and medium-sized undertakings, the viability of which is a prerequisite for the smooth operation of this system.

11.5 The State must lay down the main economic and social guidelines and draft outline provisions to be adopted by Parliament. The relevant organizations and self-governing bodies must be consulted on this matter.

11.6 In order to ensure the full development of Europe's economic power, our policy will be designed to strengthen the individual's desire to work and to encourage market competition. This calls for a policy which eliminates barriers to competition and helps to provide a more uniform basis for the activities concerned. To this end steps must be taken to harmonize, *inter alia,* taxation laws, social laws and labour laws.

12. Economic and monetary policy

12.1 Although a solution to the problems involved can only be sought at world level in the framework of international agreements, such a solution nevertheless calls, at Community level, for the immediate definition of the objectives to be pursued on a joint basis, particularly in the following fields:

- the fight for full employment, particularly for young people;

- the fight against inflation;

- the creation of a climate favourable to investment;

- the facilitation of the necessary structural changes within undertakings and at sectoral level;

- the promotion of the free movement of workers;

- the promotion of energy and research policies;

- the promotion of structural arid regional policy, accompanied by a harmonized and active social policy.

12.2   The present stage of integration favours (for internal political reasons or because of the requirements of international relations) individual action on the part of the Member States as opposed to joint action by the Community as a whole. This situation must be modified so that a genuine decision-making centre is created from which a European government will emerge.

12.3   In the longer term, it will be necessary to make Economic and Monetary Union a reality. This is one of the most important prerequisites for the maintenance, consolidation and further development of the Community's achievements.

12.4   We endorse the proposals contained in the TINDEMANS Report, which expresses the view that the achievement of Economic and Monetary Union calls for new Community efforts and adjustments:

12.4.1 • Short and medium-term economic policy must be pursued as a Community policy. This calls for the development and coordination of the areas connected with Economic and Monetary Union, i.e. of regional, social, industrial, energy, environmental and research policies.

12.4.2 • The monetary and financial policies of the Member States must be brought into line with each other. The 'snake' must be maintained as a first step towards greater stability. Steps should be taken to make it easier to extend this system to other Member States' currencies. For this reason it should once more be placed on a Community basis and supplemented by economic, monetary and financial policy measures.

12.4.3 • The powers and the size of the European Monetary Cooperation Fund must gradually be increased and brought into line with the aims of economic policy. This Fund should first be developed as an exchange equalization fund. It will guarantee a continuous adjustment of exchange rates and prevent sudden inflationary fluctuations and currency speculation.

12.4.4 • More agreement is required between the social partners. The Tripartite Conference, the Standing Committee on Employment and the Economic and Social Committee can make useful contributions in this connection.

12.4.5 • A European currency must be created.

13. Social Policy

13.1 Social policy must aim to fight against regional and social inequalities and should not simply react to the consequences of these inequalities. It must also function as a social structural policy.

13.1.1 As part of our basic philosophy, we believe that social policy should stimulate initiatives to benefit the weakest, unorganized groups of the population, in particular the handicapped and the old, and should contribute to the fight against poverty. Special attention also needs to be paid to those groups of people who in this day and age are being deported from country to country and to political refugees.

13.1.2 Social policy must contribute to the achievement of equality between men and women and in particular the realization of the principle of equal pay for equal work.

13.1.3 The Community's social policy is not intended to replace the social policies of the Member States. It is designed to stimulate the inherent powers of people, social groups and regions. It should therefore supplement national policies, provide coordination in certain cases and make for harmonization and progress. It is important to strike a fair balance between the harmonization which is essential to social progress and excessive centralization. Pluralism and versatility must be maintained in the Community.

13.2 *Partnership and solidarity*

13.2.1 Partnership and solidarity should guarantee justice for everyone and at the same time ensure that our economic system is more efficient. The Christian-Democratic option is based on freedom and justice and not on capitalism with a social face or on collectivism.

13.2.2   We attach considerable importance to regular consultations between the social partners themselves and between the social partners and the competent Community authorities.

13.2.3   We will promote at European level:

- the creation of a Charter on the rights of workers in undertakings;

- worker participation at shop-floor, plant and undertaking level, particularly by means of balanced representation on boards of directors and works councils in European limited companies;

- greater participation in property formation with a view to a more balanced distribution;

- freedom of movement and more mobility for workers based on free choice and not imposed by the social differences between favoured and less favoured regions of the Community; and

- measures for migrant workers, relating in particular to their accommodation, professional training, social, political and cultural integration and the education of their children.

13.3   *Quality of life and working conditions*

13.3.1   With a view to ensuring that the individual can experience the value of his work—and this applies to both manual and non-manual work—we are fighting for the further 'humanization' of living and working conditions, in particular by minimizing monotonous and production-line work and by reducing night work, Sunday work and shift work.

13.3.2   Community policies designed to improve industrial safety and working conditions must:

- lay down minimum Community norms;

- provide funds to enable regions and undertakings faced with particular difficulties to comply with these provisions; and

- safeguard the continuity of national efforts, but take steps to ensure that working conditions are not degraded in the process.

13.4   *Family policy*

13.4.1 Social policy must recognize, promote and safeguard the importance of the family in a free democratic society. We are in favour of a policy which enhances the cohesion of the family, strengthens its educational capacity, protects the life of unborn babies and furthers the personal development of children. A social policy with these aims must ensure that large families also have incomes which are sufficient for the upbringing and education of their children.

13.4.2 This policy must enable men and women to discharge their responsibilities in the family, at work and in society on a basis of equality.

13.4.3 The woman's job of running a home and caring for a family must be seen as comparable to other professions and valued as an important social and economic contribution to the community and the State.

14. Structural and regional policy

14.1   The European Community's structural and regional policy measures must be considerably intensified. In this connection, regional, national and Community measures must be brought together to form a coherent whole.

14.2   The Community's contribution should cover the factors listed below:

14.2.1 The expansion and diversification of the financial instruments of the Community's budget used in particular for structural measures in the following areas:

- infrastructures which are of European importance or which affect frontier regions; and

- investments, particularly in cases where it is especially urgent for inter-regional differences in capital resources and productivity to be eliminated.

14.2.2 New Community funds and the Community's lending resources, which are still largely untapped, should be used to finance these various measures.

14.2.3 The creation of Community instruments for initiatives to stimulate and encourage economic and technical development.

14.3 A large proportion of the structural problems which impede the progress of European integration concern the less developed regions of the Community.

14.4 European regional policy must be coordinated with industrial, agricultural and social policies. The funds available must be channelled to specific areas. Considerable importance attaches to private initiatives as well as to the European Regional Development Fund.

14.5 In accordance with the basic principle of subsidiarity, the Community's regional policy should not replace the regional policies of the Member States of the regions themselves. Instead it should supplement these policies and act as an instrument of coordination, guidance and stimulation. The more national and regional authorities participate in the planning, financing and implementation of measures under this policy, and the greater the responsibility they bear, the more successful the policy will be.

14.6 In this connection, special measures will have to be adopted for frontier regions.

15. Transport policy

15.1 The importance of a Community transport policy will increase in proportion to the development of a uniform Community market and the progress towards Economic and Monetary Union, which must be established on a basis of equal competition. This means that national distortions to competition in the field of goods and passenger transport must be eliminated and that a common market must be established for the transport sector and related services. As a matter of particular urgency, the Community must adopt a policy designed to counter patently unfair competition from the state-trading countries.

15.2 All forms of transport must be financially self-sufficient. To ensure that this is the case, terms of competition between the railways, road transport, sea transport and air transport must be brought into alignment and harmonized at Community level. The same applies to fiscal and social measures in this field.

15.3 Moreover, attention should be given, when working out a Community transport policy, to the contribution such a policy could make, on the basis of Community criteria, to the improvement of the results yielded by a long-term, comprehensive regional and structural policy.

16. Agricultural and fisheries policy

16.1 Agriculture is so far the only sector in which the European Community has succeeded in implementing a truly *common policy* in all the Member States. This policy should therefore be consolidated and expanded, and any undesirable features removed, on the basis of the Treaty of Rome. Efficient agriculture within the Community must also be compatible with the interests of the developing countries.

16. 1.1 In the common agricultural policy, efforts centre in particular on:

- the safeguarding of what has already been achieved, especially by eliminating monetary difficulties;

- the correction of imbalances on the agricultural market by means of an appropriate price and structural policy; this involves ensuring that the intervention system is effective in achieving a better balance, providing all producers with differentiated, but comparable, guarantees.

- the further extension of existing agricultural systems;

- the use of regional policy measures in areas with natural disadvantages; and

- the intensification of measures to improve agricultural and sales structures.

16.1.2 We shall continue to support the development of modern family farms, which have so far stood the test of the rapid evolution of our economic structures.

16.2 In its *fisheries policy,* the Community must take steps to ensure the survival and future expansion of stocks. To this end, the following four basic principles must be adopted:

- management of resources in the interest of the fishery industries of Member States;

- special arrangements for the fisheries sector in regions in which, for historical or economic reasons, this industry is of particular economic importance;

- financial solidarity in providing suitable assistance for the necessary adjustment of production; and

- political action at international level to secure maximum fishing rights in the waters of third countries.

17. Trade and industry

17.1 The European Community stands in need of a Community industrial policy. Such a policy should improve the competitiveness of European industry and create more job opportunities.

17.2 An important aim of this policy would be to give full effect to the Common Market by eliminating the remaining obstacles to trade. To this end, a legal, fiscal and financial framework will have to be created in order to make it possible to exploit fully the advantages of such a large internal market. The Community's measures must be focussed on two factors:

- an energetic policy with regard to the inevitable process of industrial reorganization, including Community assistance for the modernization and restructuring of undertakings;

- an active research policy with the aim of developing new technologies; and

- solidarity in economic relations with the countries of the Third World as a contribution to the improvement of the world economic system.

17.3   We attach particular importance to small and medium-sized undertakings, since such firms, with their will to work and their willingness to take risks, do much to improve competition and thus to contribute to a dynamic economy and an effective social system. Many of the leading figures of industry come from such undertakings. At the same time, these firms also offer workers particularly responsible positions and considerable scope for promotion.

18. Energy policy

18.1   The energy crisis has made a common energy policy essential. Such a policy must form part of the Community's foreign and external economic policy.

18.2   The aim of this policy must be to reduce oil consumption, to increase oil supplies within the Community, to diversify oil imports, to increase supplies of natural gas, to encourage coal mining in the Community, to use all energy sources sparingly, and to develop alternative energy sources and give more encouragement to the use of such sources. Where nuclear power is concerned, all necessary safety measures and precautions must be taken to protect life and the environment.

18.3   In order to reduce dependence on third countries in the energy sector, the Community must take steps to ensure, within the framework of its energy policy, that its energy requirements are covered as far as possible by its own resources. Guarantees are necessary in this field.

19. Environmental protection

19.1   High priority must be given to Community environmental policy, which must take the form of a comprehensive preventive policy. The main aims of this policy should be as follows:

- the adoption of Community minimum norms on emissions in order to prevent distortions of competition between undertakings and to afford the same protection to all citizens of Europe; and

- the attainment, in cooperation also with countries outside the Community, of common solutions to environmental problems.

20. Consumer protection

20.1   The constant changes in the range of consumer goods are making life more and more confusing for the consumer, particularly in view of the massive advertising campaigns occasionally mounted. The complexity of the markets within the Community aggravates this situation.

20.2   We therefore advocate:

- measures to protect health, in particular as regards the purchase of foodstuffs and medicines, and measures to ensure that technical equipment can be operated safely;

- objective consumer information and education;

- advertising which is consistent with fair competition and conveys the real quality of the goods and services offered; the case for introducing legally enforceable rules needs examining; and

- the harmonization of current national legal provisions in the Community.

THE COMMUNITY'S INSTITUTIONAL FRAMEWORK

The unity of Europe must be based on the determination of its peoples, which means that this unity must be truly and unequivocally democratic.

Democracy is based on the institutions which, in conformity with the law, ensure that the population can express its political wishes freely, on the formation of an authority founded on these wishes, and on the democratic control of this authority by the democratically elected representatives of the peoples.

At present, of course, national thinking and patterns of behaviour continue to impede the functioning of the Community's institutions, as do the difficult problems facing individual countries in domestic and foreign policy.

In the transitional period leading up to political union and Economic and Monetary Union, it is crucially important for the Community to move towards the establishment of a Community decision-making centre, the true partner of which will be the democratically elected European Parliament.

The direct election of the European Parliament heralds a new stage in the process of European unification. We expect the directly elected and democratically legitimated Parliament to provide a new constitutional and institutional impetus for the achievement of European union and progress towards a European federation, the ultimate political aims of unification.

Thus, the citizens of Europe, the political and social forces, the parliaments of the Member States and in particular the European Parliament, must increasingly bring their influence to bear in order to ensure that national governments adopt the decisions essential to the success of this historic enterprise.

In particular, young people must be actively involved in the construction of Europe; this applies, too, to the Community bodies which should hear the views of the responsible youth organizations on matters concerning them. The 'European Youth Forum' proposed by the Conference of Heads of State or Government in the Hague in December 1969 should be set up.

21. European Union

21.1   It is particularly important that existing Community regulations should now be made fully effective. The Community's decision-making procedures in the Council leave much to be desired. Thus, the considerable international difficulties besetting us cannot be dealt with efficiently. The Community's institutions must therefore be given more powers so that they are in a better position to cope with the requirements of authority, effectiveness, internal cohesion and double (national and Community) responsibility. The European Parliament will have to exercise a genuine influence, the Council will have to be able to take the necessary decisions and the Commission will have to be able in particular to make full use of its independent powers of initiative.

21.2   In this way we can prepare for the transition to the next phase of the process of European integration, in other words the achievement of European Union as described in the TINDEMANS report. This central aim must be achieved in the directly elected European Parliament's first term of office.

21.3   To this end,

- the existing Treaties must be applied in full and maximum use must be made of their potential;

- the authority and powers of the European Parliament and Commission, which are the main institutions representing the Community, must be strengthened;

- the powers of the European Parliament must be strengthened at least to the extent required to offset the loss of powers by the national parliaments following the transfer of responsibility to the Community;

- new powers facilitating the qualitative improvements essential to the realization of European union as the only adequate democratic answer to the challenges of our time must be transferred to the Community on the basis of proposals submitted by the Commission in agreement with the European Parliament.

21.4   The *European Council* must stimulate and encourage European unification by defining the various stages involved and in particular those new fields which are to be incorporated in Community policy and legislation. Its decisions must take the form of instructions to the organs of the Community on the implementation of such measures.

22. The Community organs

22.1 *European Parliament*

The European Parliament's powers must be extended; it should have unlimited budgetary and supervisory powers and should also be endowed with increased legislative powers. Moreover:

- every Commission should be installed in office by Parliament

- the President of the Council should report regularly to Parliament on the activities of the Council

- Parliament should have the right to take decisions instead of the Council in cases where the latter—within a given period after the conclusion of the conciliation procedure—has twice declined to take a decision on matters falling within the terms of reference of the Treaties.

22.2 *Commission*

The Commission should be the motive force behind European unification. To this end it should represent the political powers of the European Parliament more fully and become much less technocratic. This will mean that:

- the new Commission Presidents should be appointed only after consultation of the European Parliament;

- the European Parliament should hold a debate on the constitution and programme of each new Commission as it enters into office; this debate should be followed by a vote of confidence;

- since it is answerable to the European Parliament, the Commission should be placed under an obligation to adopt Parliament's proposals and to submit its own proposals to Parliament for consideration.

22.3 *Council of Ministers*

The Council must improve its effectiveness and authority as a Community institution by:

- *making its work more coherent:* as a Community institution, the Council is responsible for all areas of activity, including consultations and decisions on foreign policy in the context of 'political cooperation'; 'conciliation' between Parliament and the Council should be extended to the entire range of activities;

- *speeding up its activities:* it should return towards the practice of majority decisions on Community matters in accordance with the Treaty;

- *concentrating more on its legislative function:* it should leave detailed implementing provisions to the Commission;

- *improving the coordination of its work with that of the European Parliament:* by means of regular discussions in the European Parliament on the report of activities and by improving the procedure for conciliation between the two institutions.

22.4 *The European Court of Justice*

The European Court of Justice ensures that the Treaties establishing the European Community are fully applied in every respect; it will be the guardian of the constitution of the European Union. So that the Community can continue to function as a 'constitutional state', the Court of Justice must extend its powers to the new fields which fall within the Union's terms of reference. Individuals must be able to appeal directly to the Court of Justice against an act of one of the institutions of the Union infringing their rights.

22.5 *The European Court of Auditors*

We welcome the creation of the European Court of Auditors and hope that it will soon be fully operational. The Community's growing budget and the increasing number of cases of abuse and fraud call for the permanent supervision and strict control of Community funds. Thus, the European Court of Auditors, while fully maintaining its independence, will have to work in close collaboration with the executive institutions and the European Parliament.

23. Advisory Bodies

23.1　*The Economic and Social Committee,* which draws representatives from various economic and social circles, must increasingly assume the role of an advisory body for the definition of Community economic and social policy.

23.2　*The regions* must—through their responsible representatives—participate adequately in the adoption of decisions affecting them.

OUR GOAL: A UNITED EUROPE

For the Christian-Democrats, the European Union as described in the TINDEMANS Report and formally proclaimed by the Heads of State or Government meeting in the European Council, will represent an important step towards European unification.

We are firmly committed to the final political objective of European unification, that is the transformation of the European Union into a unique European federation of the type described, many years ago, by Robert Schuman in his declaration of 9 May 1950. This Europe will not be able to manifest its capacity for dynamic and unequivocal action until the necessary institutions have been created: a directly elected European Parliament, a Chamber of States, a European government, a Supreme Court of Justice and a Court of Auditors.

- a *Parliament,* which gives expression to the free will of the people;

- a *Chamber of States,* which represents the legitimate interests of the Member States; and

- a *Government,* which is willing and able to govern effectively.

THEN EUROPE WILL BE ABLE TO RESPOND TO ITS VOCATION.

Basic Programme, Adopted by the Ninth EPP Congress in Athens on 12–14 November 1992

BASIC PROGRAMME
adopted by the IXth EPP Congress, Athens, November 1992

Contents

P R O L O G U E

The new European society

Europe is undergoing a period of considerable change. Its States and populations are having to adjust to new circumstances. European society is seeking a new identity. The end of ideological, political and military confrontation between East and West has created new openings as well as new opportunities for agreement and cooperation.

For us Christian Democrats, members of the European People's Party, this time of upheaval offers us a unique and unprecedented task. Our prime mission is to seize those opportunities offered and use them to the full. Periods of profound change also generate imponderable issues, dangers and potential for conflict.

Furthermore, our democracies are experiencing a profound crisis of values challenging the political system right down to its foundations. We Christian Democrats are aware of these dangers, temptations and the loss of direction resulting from this crisis. We cannot provide perfect solutions to overcome them but we act on the basis of fundamental values and principles which point us in the right direction.

Against ideological temptations

Regardless of the disappearance of Marxism-Leninism as a foundation for society in Central and East Europe, the end of ideologies is not really in sight. Rather, competition between ideologies is occurring at other levels.

At first sight, liberal ideology has many advantages. The market economy has allowed the development of living standards other systems have not yet been able to achieve. Neoliberalism, however, ignores the social dimensions of the free market economy by unilaterally stressing the individual efforts of each man and woman; which can only work against the weakest members of society. Once again, this leads to conflict and confrontation, affecting solidarity which must also be a valid part of the international context.

Ecology provides a positive contribution, striving to call upon the "best" in every man and woman to advance the quality of life. Within this ideology, however, lies the temptation to giving absolute values to nature and the earth, opposed to all technological and economic progress, resulting in the limitation of freedom and self-fulfilment of mankind, which consequently does not contribute to respect for Creation.

The most dangerous response to the fall of communism is a misplaced nationalist ideology. The feeling of patriotism and of belonging to a specific community is inherent to the existence of any human being and thus entirely legitimate. Nationalism can no longer be acceptable, however, when it becomes absolute and denies other values and responsibilities such as loyalty to the democratic state and the rights of minorities.

At first sight, socialism would seem to be the natural heir to communism. It is nonetheless also handicapped; either it uses the old model of class struggle and class opposition, a model which no longer satisfies its citizens' aspirations, or - as is the case in social democracy - it gives up the class fight but remains suspicious of civil society and intermediary bodies and gives too much priority to the activity of the State, thus all too often invading the social fabric.

We Christian Democrats see the weaknesses in these ideologies which are bound to mislead us in the end. We also reject a purely pragmatic approach to problems in society. We wish to impart a valid response to those aspiring to a more human society. Our society cannot be restricted to satisfying material needs alone. The needs of human beings in their totality and and their fulfilment in the framework of a completely new society are crucial for the realization of their wellbeing.

Overcoming new threats

European society sees itself confronted by multiple contradictory developments jeopardizing its internal cohesion.

Never before have demographic developments reached the scale of the challenge that will exist over the coming decades. Major migratory movements in the world result from overpopulation in certain regions and the attraction of developed, stable countries.

The gap between rich and poor cannot remain one of the world's greatest tragedies for much longer. European society faces the additional problem of ageing which has profound implications in terms of the organisation of society and of the need for practical expressions of solidarity and concern for others.

Economic globalization is leading to growing interdependence. But rules are still lacking for these complex and reciprocal relations which must be organized in a coherent manner. We must be watchful of the draining of natural resources through misuse and the danger of ecological catastrophes.

Scientific knowledge is a potential source of well-being and of fair distribution of prosperity. Technological innovations will help improve the quality of tomorrow's society. Nevertheless, technological and scientific developments must not be allowed to misguide man or cause him to lose his respect for nature and its limitations.

The meaningless flow of information, especially when deformed by the media, may result in the total disappearance of responsibility and the invasion of irrationality into politics. It can create a climate of instability and confusion as well as leading to apathy and the impression that there are quick and easy solutions to all problems.

The basic elements for consensus achieved in European society are being threatened by a rising wave of racism and nationalism as well as a rebirth of criminality and aggressiveness.

We Christian Democrats do not claim to have ready-made, convincing solutions to all these evolutions. Without losing sight of the threats which exist, we intend to act on the basis of the signs of hope that are appearing in our society today.

An end to the East-West conflict has opened up opportunities for lasting world peace and savings on defence expenditure so that more money can be dedicated to developing populations.

A consensus for solutions to many of the world's problems can henceforth be more easily found through international organizations, as well as the creation of a juridical world order. We Christian Democrats rejoice in the irresistible rise of parliamentary democracy. The model for western consensus linking social market economy to democracy has achieved appreciable success.

The explicit acceptance which this combination of political and economic freedom is obtaining throughout the world is very encouraging and acts as a decisive stimulus for us to remain faithful to this model.

Understanding the signs of the times

Perceptions of the values of European citizens differ and often appear contradictory: a withdrawal into the private sphere; the growth of materialism, individualism and cynicism, together with the collapse of universal ideals; at the same time a commitment towards specific issues in society (single-issue politics) but also an aspiration to lead one's life on the basis of values such as responsibility, loyalty and a concern for security.

We Christian Democrats, members of the European People's Party, wish to make a positive contribution to these changes in the spiritual climate and seize these opportunities. Many citizens, whether adhering to a religion or Church or not, are willing to collaborate, to be committed and to demonstrate solidarity. Christian Democracy, on the basis of its political tradition, seeks to appeal to what is "best", to the "constructive" aspect which exists in each human individual, and to give contemporary expression to the ideals of social Christian personalism.

CHAPTER I

FOUNDATIONS AND REFERENCE VALUES

Our concept of man

101. We Christian Democrats, members of the EPP, affirm the inalienable dignity of every human being. We regard man as the subject and not the object of history.

102. On the basis of Judaeo-Christian values, we regard every man and every woman as a person, i.e. as a unique human being who is irreplaceable, totally irreducible, free by nature and open to transcendence.

103. Each human being within society depends on others. Because they are free, responsible and interdependent, people must take part in the construction of society. For many of us, what lies behind this commitment is the belief that we are called on to contribute to God's work of creation and freedom.

104. Freedom is inherent in the essential nature of man. It means that every individual has the right and the duty to be fully responsible for himself and his acts and to share responsibility vis-à-vis his neighbour and creation.

105. Everything leads us to affirm that truth is transcendent and as such is not entirely accessible to man. Our concept of freedom leads us to affirm that man is by his nature fallible. Consequently, we acknowledge that it is impossible for anyone to conceive of far less construct - a perfect society, free of all pain or conflict. We reject any form of totalitarianism based on such an aspiration.

106. In accordance with our concept of man, we affirm that all men and all women have the same dignity and are by their nature equal.

Fundamental values

107. We affirm that every woman and every man is responsible for constantly improving the society in which they live on the basis of reference values and regulating principles defined in common. It is by applying these that they will be able to prevent, manage and settle peacefully the differences between men and the problems and challenges facing mankind.

108. Christian Democratic thought and political action are based on fundamental, interdependent, equally important and universally applicable values: freedom and responsibility, fundamental equality, justice and solidarity.

Freedom and responsibility

109. We believe that genuine freedom means autonomy and responsibility, not irresponsible independence. It renders every person responsible for their actions according to their conscience and also before their community and the future generations.

110. According to this conception, everyone shares responsibility vis-à-vis the created world. Future generations must also be able to live in harmony with a natural environment in which each human being is an essential link in the chain. Groups, communities, peoples, nations and states are therefore answerable for their actions to each living and future human being.

111. True justice and solidarity cannot exist unless prior to this the existence of freedom is accepted by everyone as an essential condition.

112. That is why we Christian Democrats want to enable everyone to enjoy in their daily existence the inalienable rights recognized as belonging to every man and woman and their communities. This involves both the guarantee of the right to develop and use their gifts, talents and abilities to the full, and on the other hand the obligation to offer these in service to the community and to seek at all times to apply the values of justice and solidarity in relationships with others.

113. This freedom, that is at once the condition and the consequence of the constant endeavour to apply the values of justice and solidarity, also applies to the existing authorities, both in their internal organization and in their relationship to private individuals. This also has certain implications for our Christian Democratic concept of the political system.

114. The authorities derive their legitimacy from the requirement to establish the appropriate conditions for the personal development of each and everyone on a community basis. Any authority, whether public or private and at whatever level it operates, must therefore under all circumstances protect the general interest and the common good.

115. The general interest and the common good must not be confused with the sum of individual interests. However, they must always be compatible

with a proper respect for the ensemble of individual, civil and political, economic and social, cultural and collective rights of each person.

116. The right of peoples to self-determination and the free exercise of their legitimate rights cannot be invoked to deny any one person the exercise and enjoyment of his or her rights. This right of peoples is nevertheless a high form of justice given that it is the affirmation and recognition of a sense of identity and the wish to live together in freely chosen politico-social structures.

Fundamental equality

117. All human beings are equal because they are endowed with the same dignity. In relationships with others, each person's freedom is therefore limited by a respect for others' freedom deriving from the recognition of that fundamental equality.

118. Notwithstanding their differences in terms of gifts, talents and abilities, each person must be able to achieve personal development in freedom and equality at his or her own level, whatever his or her origin, sex, age, race, nationality, religion, conviction, social status or state of health.

119. The same rights must be recognized and the same duties imposed according to each person's capabilities.

Justice

120. The concept of justice means that the necessary conditions for individuals and also their communities - depending on their nature and objectives - to exercise their freedom must be guaranteed at all times. It is the characteristic of justice to attribute to each individual his due, and actively to seek out greater equality of opportunity and a life in society which is based on harmonious relations.

121. One dimension of justice is respect for the law. Laws are constantly evolving in tune with the dynamics of civilization and technical progress, but they must always have been freely accepted by men and their communities.

122. Laws must evolve on the basis of universal respect for man's fundamental and inalienable rights, as defined in the 1948 Universal Declaration of the Rights of Man and the 1950 European Convention on Human Rights and Fundamental Freedoms. These declarations enshrining individual rights (first generation) and social rights (second generation) should be supplemented by a declaration on a third generation of fundamental rights, such as the right to information, to an unpolluted environment, to privacy and to genetic identity. All such rights must be enjoyed by both individuals and communities.

123. Justice cannot be arbitrary or confused with the dictatorship of the majority. It requires respect for the minority, to whom no majority can deny the free exercise of its rights. Justice cannot, moreover, be incompatible with fundamental rights and freedoms, which must also be taken equally into consideration in the application of law.

Solidarity

124. Justice, the fundamental equality of all men and the inalienable dignity of each individual are indissolubly linked to a spirit of solidarity. It constitutes an essential component in the establishment and deepening of more humane relations between men and between their different communities as well as within them.

125. Solidarity means an awareness of the interdependence and interrelatedness of individuals and their communities. It also means practical action, sharing, effective aid, and rights and duties in relation to individuals and their communities which form part of a whole and in the final analysis fuse into the universal. Consequently, anything that happens to one person has repercussions for others.

126. For Christian Democrats, solidarity means above all protecting those who are weakest in our own society and in the world.

127. In affirming the unity of mankind in time and space, we see solidarity as not only horizontal, between living beings of all generations and all places. It is also vertical, extending to a consideration of the legitimate interests of future generations and including respect for the created world.

The implementation of values

128. Our concept of the person rejects both selfish individualism and collectivism as a reducing factor. Each person belongs to a community and must subordinate his or her individual interests to the legitimate authority of the community by accepting the constraints necessary for the protection of the fundamental rights and freedoms of its other members.

129. The person is also the end of every community since the source of legitimacy of any power lies precisely in its attempts to ensure the personal development of all those subordinate to it.

130. Convinced of the inalienable dignity of man and the freedom and equality of all, we reject extremes and advocate dialogue. We reject exclusion and advocate tolerance and sharing. We want to see all people enjoying autonomy while respecting other peoples and communities and the personal convictions of each of their nationals.

131. In short, the Christian Democratic concept of man and of society focuses on the integral development of every individual in a way that satisfies his material, cultural and spiritual needs, whilst at the same time respecting the freedom of others.

132. Finally, we reject any attempt to systematize thinking within a closed or dogmatic mould.

Respect for the created world

133. We oppose the unthinking and unjust exploitation of the earth, without respect for the self-regenerating potential of nature. Our concept of man calls for management of the earth with a view to satisfying the needs of all and improving the living conditions and quality of life of everyone, while also ensuring lasting development compatible with the protection of the legitimate interests of future generations.

134. Respect for the created world means that responsible management of the biosphere and forms of life which make up mankind's common heritage is both necessary and essential to the harmonious development of every living and future human being.

135. The developments under way in all spheres of society must not impede the potential or deplete the resources of future generations. This concept of sustainable development is bound up˙with that of responsible growth and must be incorporated into every policy, at whatever level of power.

136. Sustainable development means inter alia reconciling the requirements of the economy with those of the environment and taking account of the protection of the environment and nature when taking any economic, social or political decisions. We must act in a way that ensures that all potential is safeguarded for future generations.

Our vision of society

137. Our values must be applied not only in the political sphere but also in the economic, social and cultural spheres.

138. Economic development, based on the contributions of each and every person, cannot foster well-being or peace unless its fruits are equally distributed, with the aim of improving the living conditions of each person and his or her personal development. It is for this reason that we must support and develop systems of collective solidarity.

139. It is necessary to remain vigilant with regard to the danger of economic power being misused as an instrument of domination or injustice. It is therefore important to safeguard competition between market forces.

140. Finally, it is vital to try to ensure social justice and solidarity based on partnership and participation at all levels - private, national and international.

141. On the basis of these values, the Christian Democratic vision of society is based on the principle of subsidiarity.

142. The principle of subsidiarity means that power must be exercised at the level which corresponds to the requirements of solidarity, effectiveness and the participation of citizens, in other words where it is both most effective and closest to the individual. Tasks that can be performed at a lower level must not be transferred to a higher level. However, the principle of

subsidiarity also means that the higher level must help the lower level in the performance of its tasks.

143. This principle is based on the premise that society can be constructed in freedom. The public authorities must therefore respect human rights and fundamental freedoms, recognize the relative autonomy of social groups and not take the place of private initiative unless the latter is weak or non-existent.

144. In accordance with the principle of subsidiarity, the European People's Party advocates the creation and strengthening of intra- and international conventions and bodies where they are more capable of providing a joint response to problems.

145. In this same spirit, the European People's Party encourages the activities of nongovernmental organizations and the creation of associations of every form and latitude.

146. The general application of the principle of subsidiarity allows for the permanent recognition of the particularities and specific characteristics of each person and each community whilst affirming that they form part of the universal community of mankind.

147. It encourages awareness of the need for a genuine international partnership seeking the common management of the planet and its common heritage, on the basis of respect for the irreducible differences that exist between individuals, communities, groups, peoples, nations and states.

148. The increasingly cooperative nature of this management has become all the more vital in view of the fact that the techniques that man has put in his service over the past two centuries in order to dominate nature for his profit are now putting nature - and consequently also the survival of mankind - at risk.

149. This exponential development of technical resources has also led to a general awareness of the need for a genuine international partnership aimed at the common management of the planet.

150. At the same time, this exponential development necessitates an in-depth review of the concepts of unlimited economic growth and a purely material quality of life with a view to achieving an enduring development that responds to the needs of today without endangering the living conditions of future generations or the satisfaction of their basic needs.

151. Respect for the principle of responsibility and autonomy henceforth requires man to exercise his powers with self-restraint at every level.

Our concept of the political system

152. We consider that democracy is a vital condition for the development of individuals.

153. Our commitment to the development of individuals implies developing and strengthening everywhere the constitutional state in order to prevent the relationships of violence which are still found all too frequently not only between individuals but also between communities.

154. We consider that there is no alternative to democracy but that it must nevertheless be adapted to different cultural and socio-economic situations on the basis of the respect for a universal framework defined by human rights and fundamental freedoms.

155. The participation of each person in public life and in decisions that concern him or her represents an essential element of democracy.

156. Our expressed resolve to counterbalance the principle of subsidiarity with the recognition of diversity and international partnership with the participation of each person in public life, reflects a search for harmony in the framework of a constitutional state in which the common laws vital to all life in society may be defined and applied on a basis of respect for the inalienable rights and freedom of all.

157. Applying the principles of subsidiarity, international partnership, participation by all in public life (especially through free elections held at regular intervals, based on secret ballots and universal suffrage) and the constitutional state enables each person to achieve personal development based on respect for others and progress to be made towards the resolution of conflicts.

158. The limits imposed by the principle of subsidiarity also contribute to a specific separation of powers by preventing their concentration.

159. Every authority is in effect at the service of the individual. No state can therefore use the pretext of respect for its sovereignty in order to violate the rights and fundamental freedoms of persons or communities.

160. If it does, the international community of states must take protective measures, on the basis of treaties, conventions, agreements and other texts, and even by codifying a graduated obligation to intervene, subject to strict international, democratic control.

161. The raison d'être of the sovereignty of states is to enable them to work freely and as best they can to ensure the well-being and development of their people and to defend and reinstate international juridical order. This also means, however, that states must share their sovereignty in supranational and international organizations where they cannot take effective action individually.

162. The European People's Party wants to help build a world that is based on freedom and solidarity, in which every man and every woman is viewed as a human being in all his or her fullness and complexity.

An appeal to values

163. As Christian Democrats we stress the need to distinguish between the roles of the Church and the State in society, between religion and politics. However, we reaffirm the link that exists between, on the one hand, Christian values based on the Gospel and Christian cultural heritage and, on the other hand, the democratic ideals of freedom, fundamental equality between men, social justice and solidarity.

164. These principles and values lie at the heart of the European People's Party's political thinking and action. We derive our strength and our motivation from a constant reference to our values.

165. As a Christian Democratic but non-confessional party, the European People's Party is essentially a political party of values.

166. If it rejects, forgets, neglects or dilutes its values, the European People's Party will be no more than an instrument of power, without soul or future, while also forfeiting the universal and original nature of its message, which is based on a global apprehension of the irreducible complexity of every human being and of life in society.

CHAPTER II

FROM THE EUROPEAN COMMUNITY TO THE EUROPEAN UNION

201. Only the union of Europe can secure its future: a future of freedom and security, progress and solidarity. In line with the commitment to Europe which the Christian Democrats have shown since the very beginning, the EPP calls for a gradual - but resolute - transformation of the European Community into a genuine political union on a federal model, following the doctrinal lines defined by the congresses of Luxembourg in 1988 and Dublin in 1990.

For a federal Europe

202. A federal Europe is now more than ever a necessary and realistic political objective. It is necessary because the radical changes occurring on the European continent must take place within a structured, democratic and peaceful framework. Only a federal organization of Europe can match the aspirations and interests of Europeans who want to share a common destiny. It is realistic because history is speeding up and people are ready for an acceleration of the process of union based on delegating and sharing national sovereignty.

203. The EPP considers that only a federal construction of the European Union can: - on the hand, guarantee unity within diversity and hence respect the national identities and cultural and regional diversities that characterize Europe and result from its history; - on the other hand, ensure a common approach to the solution of common problems.

204. In the modern world there are few crucial questions that are confined to a purely national context. If the Member States want to carry out effec-

tively the national tasks for which they are responsible, it will become increasingly urgent for them to find European solutions. That is why the Community patrimony must be preserved and developed.

205. The European Union must be founded on a relationship of federation and not subordination between it and its Member States. This federal concept will take account of local, regional, national and European levels. The European Federation will be a community of decentralized nations, not a unitary super-state.

206. The distribution of powers between the Union and the Member States and the regions must be organized on the basis of the principle of subsidiarity, which means that any action by the Union will be subsidiary to action by the states and regions. The Union must therefore be granted only those powers of which it can make the best use. In other words, the Union will have competence in the areas where it can act more effectively than the Member States could individually, because the scale or effects of the actions involved go beyond national frontiers.

207. The principle of subsidiarity must henceforth be expressly applied to the activities of all the Community institutions. The national, regional and local authorities will retain their specific role and function in this context. Obviously, specifically national, regional and local powers, and the diversity which results from them, must be respected.

208. But although the states and the regions must retain sufficient and adequate autonomy, it is equally important that the Union is not subordinate either to the states or to the regions in areas where it must act in the general interest.

209. As in any federal-type system, it will be necessary to distinguish clearly between the exclusive powers of the Member States, concurrent or shared powers and the exclusive powers of the European Union, it being understood that the distribution of exclusive powers and concurrent powers may change.

210. Both exclusive and concurrent powers must be carefully defined and the Union will have only those powers which are expressly allocated to it, all other areas remaining within the power of the states or regions.

211. The Constitution of the Union will have to establish effective mechanisms and procedures for allocating areas of competence not foreseen when it entered into force. These new areas will be necessary in order to ensure that the Union remains capable of adapting to new economic, social and technological challenges and to the needs of European development and the international political situation.

212. The Union must be given all the means necessary for the achievement of its objectives and the implementation of its policies. It will therefore be given a federal-type budget with sufficient resources managed on a 'progressive' basis, taking into account the relative prosperity of each Member State.

213. In this connection, the EPP is in favour of a direct relationship between the European Community and the taxpayer, thereby also giving the European Parliament direct responsibility vis-à-vis the taxpayer. However, the financing of the European Community must take into account the financial situation of both the Member States and the Community. The attribution of fiscal powers to the Community, as provided for in the Draft Treaty on European Union adopted by the European Parliament in 1984, must not lead to an increase in the overall tax burden.

214. The EPP's institutional programme draws on the advances - and the gaps - in the Maastricht Treaty of 7 February 1992, which sanctioned the transition from the European Economic Community (EEC) to the European Community (EC) and to the European Union.

215. The Maastricht Treaty is an important step towards European Union. Its potential must be fully exploited and its shortcomings rectified.

216. Pending the future transformation of the Community into a genuine European Union, the Maastricht Treaty of 7 February 1992 has ratified a complex structure incorporating different institutional models.

217. The structure adopted by the Maastricht Treaty comprises three different "pillars".
    1. The first, of a federal type, is the actual Community legislative heritage (EEC, ECSC, Euratom), reinforced by Economic and Monetary Union and certain new powers.

2. A second pillar, that of judicial and home affairs cooperation, is essentially intergovernmental, although some 'bridges' have been developed linking it with the Community structure. However, the Court of Justice has been expressly excluded and there is no real parliamentary control even though it is a field directly involving the rights of citizens.

3. The third pillar - common foreign and security (and, in time, defence policy remain essentially intergovernmental. But the proposed 'bridges' and other in terim ad hoc procedures intended to ensure the cohesion o f external policy activities may be used to promote t he process of 'communitarization' by osmosis. The development of joint action in the sphere of foreign policy will require an extension of majority voting.

218. The EPP will remain watchful that intergovernmental action does not eventually take over from Community action. In particular it will ensure that the review of the Maastricht Treaty, which it hopes to see happen before 1996, will lead towards greater communitarization and restore the unitary nature of the draft treaty adopted by the European Parliament.

219. The EPP strongly reaffirms the federative vision of the Christian Democratic pioneers of Europe. It stresses that the federal goal of European integration must be explicitly included in the Treaty on European Union.

For an effective Europe

220. The future of Community integration will depend on the Community's (or the Union's) actual ability to anticipate and resolve the specific problems of concern to its citizens.

221. The climate of uncertainty about the new European order following the collapse of the Communist regimes makes it more important than ever to strengthen the only existing stronghold, namely the Community, in its decision-making capacity and capacity for action both within and outside.

222. The Single European Act of 1986 marked a major although limited step forward in the decision-making process by providing for qualified majority voting on a dozen or so matters connected with the 1992 objective.

This resulted, among others, in the measures relating to freedom of capital movements, control of mergers, recognition of degrees and the opening of public contracts.

223. But the most important measures remained blocked:

1 - the objective of the free movement of persons ;

2 - other objectives requiring Council unanimity, such as the total abolition of fiscal frontiers, were postponed;

3 - in some fields , such as energy, telecommunications, transport, postal services and payment systems, the internal market will still not be completed on 1st January 1993.

224. It is clear, however, that in order to attain the already established objective of the internal market - the central axis of the Single Act -the qualified majority vote has to apply to sectors formerly requiring unanimity (taxation, free movement of persons, etc). This is even more true of the new objectives which the Twelve set themselves in the Maastricht Treaty.

225. The progress of the Community towards a genuine European Union therefore implies above all an institutional system that is able to assume its responsibilities effectively.

226. First, that means meeting the new commitments entered into in the Maastricht Treaty, such as:

1 - increased protection for the rights of European citizens, common policies in fields such as immigration, right of asylum and help for refugees and effective combating of crossborder crime and terrorism at European level;

2 - formulating a common foreign and security policy, eventually to include a common defence policy;

3 - completing a single market on the basis of common policies and guaranteeing economic and social cohesion, growth based on respect for the environment and a high level of employment;

4 - the creation of an Economic and Monetary Union on the basis of a single currency and an autonomous central bank, together with respect for the procedures and timetables set out to that end;

5 - the obligation to tackle the new Community activities decided upon in the field of social policy, energy, civil protection and tourism;

6 - the extension of the scale of Community powers in the fields of consumer protection, public health, research and development, industry, trans-European networks (transport, telecommunications, energy) and the European dimension of culture and education.

7 - lastly, respect for the principle of solidarity between the Member States.

227. Secondly, the Community must remain capable of achieving the objectives it has set itself even in the event of the accession of new Member States. Strengthening the Community institutions is in effect a necessary - though not sufficient - condition for the success of its future enlargement.

228. A debate and general decision on the institutional changes required by Community enlargement should precede rather than follow accession negotiations.

229. Decision-making - or rather co-decision-making - procedures must be substantially improved, inter alia by holding meetings of the Council of Ministers in public when they involve legislation and ensuring that they are subject to democratic control by the national parliaments.

230. The Council procedure of unanimous voting must gradually be restricted. First of all, the field of application of the qualified majority vote must be extended to areas of prime interest to the Community, such as important aspects of social and environmental policy.

231. The Commission is the Community's engine. From now onwards, its composition and powers will be more closely adapted to the needs of efficient management, taking account of the principle of subsidiarity and, above all, future enlargement. The EPP is therefore in favour of the emergence of a genuine European executive power, independent of the Council, which will hold legislative power together with the European Parliament and become a Chamber of States.

232. The distribution of powers must ensure that excessive technicalities do not impede the functioning of the Community (or the Union). The latter must deal only with clearly-defined, essential issues. The Member States and their components (regions, Länder, etc.) must ensure that Community laws and decisions are applied fairly.

233. The Court of Justice, which interprets and ensures observance of Community law, will have to be given the right to impose sanctions on Member States that do not respect its decrees.

234. The Community should be able to take executive measures directly in the event of refusal to act or even a passive attitude by the national authorities within a reasonable period of time.

For a democratic Europe

235. Further efforts must be made to give the European Community genuine and fundamental democratic legitimacy. The commitment to democratic ideals that is shared by all the Member States - and required of all applicants for accession - must form an integral part of the Community system in general and its decision-making process in particular.

236. It is the European Parliament, elected by universal suffrage, which primarily ensures that Europe is built on a basis of democratic legitimacy. It must therefore have the final say on constitutional and legislative matters.

- The requirement of the European Parliament's assent must be extended to new actions by the Community (Article 235 of the Treaty of Rome) and to the revision of the Treaties (Article N of the Maastricht Treaty).

- The co-decision procedure must be simplified and apply to all laws of a general scope and throughout the legislative procedure; the Council may not decide unilaterally in the event that it does not obtain Parliament's agreement.

237. The increase in Parliament's powers must not, however, be at the expense of the Commission. Having been given greater democratic legitimacy (appointment by Parliament), the Commission must now continue

to exercise its right of initiative at every stage of the Parliament-Council legislative process.

238. The national parliaments must be more closely associated with the common endeavour through the creation in all the Member State parliaments of a committee on European affairs, debates on European issues in each parliamentary session and good cooperation with the European Parliament.

For a social market economy that respects the environment

239. As a result of their market economy structures and social guarantees, the Community Member States have achieved a better balance in industrial relations than other economic and social systems, as can be seen from the concrete achievements of social justice, social progress and social guarantees for individuals. We Christian Democrats wish to uphold the principle of the market economy and strengthen the social balance in Europe.

240. Our national economies, based on market economy structures, must be successful while at the same time taking into account the social dimension and the environment.

241. Europe will have no meaning unless it is both an economic and a social Europe. At present there is an imbalance in this respect. The social deficit therefore needs to be made up and internal cohesion strengthened as the single market is completed and Economic and Monetary Union takes shape.

242. Since the establishment of the ECSC in 1951 and the EEC in 1957, the "de facto solidarity" relating to vital economic interests has laid the foundations of an "ever closer union among the peoples of Europe". The present stage of Economic and Monetary Union is based on the convergence of the economies of the Twelve and the definition of a single monetary policy. The inflation rates, deficits, tax policies, etc. of the Member States must converge closely enough to enable them to unite naturally and end up by adopting a single currency. The dates and constraints must be fixed firmly and respected in order to mark out the route.

243. However, the economy is not an end in itself: it is a means at the service of a concept of society based on the individual human being, that is to

say on freedoms and necessary solidarity. To build Europe is therefore more than a mere economic project.

244. In any case the Community's field of action has been gradually extended to cover matters that are not strictly economic. The EPP has advocated and supported that process.

245. The EPP's European policy is based on a coherent conception of society, every aspect of which must form part of a balanced progression towards European Union. This choice of society is based on the Christian-Democratic principles of freedom, fundamental equality, justice and solidarity (including attempts to overcome regional inequalities), social dialogue, respect for cultural differences, a social market economy, protection of the environment and openness towards the world.

246. The EPP points out that to reduce the European ideal to economic imperatives would in the end merely multiply the constraints on solidarity. That is why Christian Democrats are in favour of a social market economy that respects the environment.

247. In line with this concept of social solidarity - which is specifically Christian-Democratic - Economic and Monetary Union must obey the rules of public interest and social justice.

248. We must guarantee that the single market is completed on an economically and socially balanced basis and that the social and environmental dimensions of Economic and Monetary Union develop harmoniously. We must also guarantee that the process of unification is based on solidarity -defined as economic and social cohesion - between the Member States and regions of Europe and on the principle of subsidiarity.

249. The objective of the social market economy must be to strengthen economic, monetary and social cooperation even further and ensure that the citizens of Europe enjoy equal opportunities, greater prosperity, better environmental protection and social progress.

250. With respect to the environment, we Christian Democrats undertake to preserve and restore the foundations of life in Europe and the world. We expressly pledge to preserve the earth for both ourselves and our children.

251. The EPP considers that environmental protection is one of the greatest challenges of the 1990s along with the preservation and development of the European social system. We can no longer allow the growing costs resulting from the impoverishment and destruction of the environment to be borne by everyone - i.e. by the state - while those responsible for pollution continue to derive profit from their socially detrimental actions.

> 1 - We want to incorporate the protection and restoration of the environment into the market economy system, unlike those who speak of an insoluble conflict between the interests of the economy and those of the environment.

> 2 - We regard the challenge of protecting the environment as a fascinating task, unlike those who are pessimistic and hostile to progress.

> 3 - We want to use the dynamism of qualitative growth to help protect the environment, unlike those who are aggressively hostile to growth and in fact reject our free society system.

We trust in solutions which go hand in hand with the market economy and with man's intelligence and desire to learn, unlike those who are content with spectacular pseudo-solutions or who call for increased public sector activity, more bureaucratic planning or more direct state intervention.

252. The EPP considers that the basic elements of European social policy are as follows:

> 1- the improvement of living and working conditions through the provision of an appropriate income and measures to provide jobs for all those willing to work, the creation of new jobs and measures to combat unemployment, recognition of the right to cooperation and responsibility in the economy;
>
> 2 - the promotion and development of vocational training and further training and retraining measures with a view to reintegrating all those who have been excluded from the employment market and to ensuring that workers can meet the new challenges of the modern job market;

3 - equal opportunities and equal rights for men and women: every policy in this sphere must aim to combat all forms of direct and indirect discrimination are all too frequently found in various sectors of activity. This principle can only be put into practice in a socio-economic context which makes it possible to reconcile family life with working life;

4 - effective protection against racial discrimination;

5 - the free movement and free provision of services of workers and citizens;

6 - reintegration into social life of the handicapped and of victims of extreme poverty;

7 - the possibility of giving legal force to European collective agreements;

8 - the reduction of the disparities in social security cover in the various Community Member States;

9 - the reorganization of working time to enable men and women to reconcile family commitments with working and social life;

10 - the promotion of a society favourable to children and families;

11 - the integration and participation of the elderly in a society based on the principle of solidarity;

12 - development of rural areas.

253. The future constitution of the Union must also encompass areas of social policy which, in accordance with the subsidiarity principle, cannot be dealt with by the Member States alone. The implementation and extension of the Social Charter, which has been more or less blocked by the unanimity rule, must be guaranteed. It is also necessary to promote dialogue between the social partners and work towards the emergence of collective agreements alongside or in place of conventional regulations.

For a Europe open towards the other Europe

254. The Community needs to be able to take stronger and more united external action in order finally to secure - or acquire - credibility on the international political scene.

255. Having served as a model of reconciliation and prosperity for the entire European continent for forty years, the Community must naturally be

involved in the forefront of the democratic changes in Central and Eastern Europe. The future enlargement of the Community to include some members of the European Free Trade Association (EFTA) will, moreover, help the Community in the accomplishment of this task.

256. The disappearance of the "pax sovietica" reflected the beginning of a promising process but one that was also lengthy, difficult and dangerous for the people of that part of our continent.

257. The political readjustments in Central and Eastern Europe and the ex-Soviet Union will lead to the establishment of a new European security system. It will have to be constructed in such a way as not to exclude any nation and to respect every nation, with a view to promoting international peace and security and condemning the use of force (or the threat of the use of force) by any one country against another.

258. The transition from planned economies to open market economies throughout the world is a difficult one. In effect, it calls for a radical transformation not only of economic techniques and management procedures but also of ways of life and attitudes.

259. The European Community must therefore continue and redouble its efforts to ensure that these Central and Eastern European countries have the human and material resources they need to carry out the necessary transition - peacefully and at the lowest possible social and cultural cost - they have undertaken.

260. Another major objective of the European Community must be to encourage the establishment in Central and Eastern Europe of constitutional states that respect the individual, economic, social, cultural and collective rights of their nationals. These states will be fully integrated into the community of free democracies in Europe and the world, on the basis of strict equality between partners.

261. The European Community will also have to encourage the various Central, Eastern and south-eastern European countries to step up their trade with the West and among themselves, on the basis of regional agreements that are, where possible, multilateral.

262. At the same time as the EC states pursue economic and political integration, all the European nations together must equip themselves with the structures and resources needed to create deeper, enduring and mutually profitable pan-European cooperation on all international questions (especially those covered by the Helsinki Final Act).

263. The Community must remain open towards any other European states that may wish to share the common destiny of our twelve countries, provided that they share the values of Europe and genuinely practise multi-party parliamentary democracy and provided also that they have a social market economy compatible with ours and that their accession to the Community will not be a threat to this. Moreover, countries wishing to join the Community must accept the Maastricht Treaty and be prepared to take part in all the plans laid down therein (the internal market, Economic and Monetary Union, common foreign and security policy and cooperation in justice and home affairs).

264. The countries of the "other Europe" belong to Europe. So they must certainly "find their way back into Europe" by gradually becoming integrated into the European Community. The Community, for its part, will have to prepare itself for this, which means above all consolidating the Community, further developing the European Economic Area (EEA) and gradually consolidating the cooperation and association agreements with the countries of Central and Eastern Europe.

265. Any future enlargement of the Community presupposes the following:

- the gradual realization of the political and economic conditions necessary for those states that so wish to become members;
- the creation of a network of relations designed to speed up the process of cooperation and integration in a new European order.

266. To that end, the Council of Europe and the CSCE will extend and adapt their activities, and the European Treaties and the cooperation and association agreements will also be extended and adapted in order to meet the requirements of a creative, generous and cooperative policy.

For a security and defence policy in Europe

267. The end of the East-West confrontation considerably reduced the risks of large-scale armed conflict in Europe. However, the survival in the Commonwealth of Independent States (CIS) of the former Soviet Union and in neighbouring states of forces and institutions opposed to change constitutes an enduring risk to European security. Moreover, the collapse of the Soviet empire has led to a resurgence of national and ethnic conflicts, as in the former Yugoslavia, which may also explode elsewhere in Eastern Europe. Lastly, European security requires effective political control over the arsenals of nuclear and chemical weapons that are still scattered around the CIS.

268. At the same time, Europe remains vulnerable to the repercussions of what is happening in the Middle East and North Africa. The population growth in these regions, together with the growing attraction of fundamentalism (and the intrinsic problems this generates), could give rise to major instability in the region which would directly affect Europe.

269. Under these conditions, Western security structures such as NATO and the WEU have an important role to play in the system of European stability. Appropriate ways must be found to link up the Central and Eastern European states with the Western security systems in the medium term.

270. The aim of the WEU, which forms an integral part of the European unification process, is to define, in cooperation with NATO, a common defence policy that will lead as soon as possible and by 1998 at the latest to "common defence" in the framework of the European Union when the WEU Treaty is renewed after fifty years.

271. The creation of European structures within the WEU and their institutionalized cooperation with those of NATO must begin immediately. The missions of the European armed forces must be defined and provision made for the following tasks: common defence of the allies, the guarantee of independence and territorial integrity of the European Community, maintenance and restoration of peace, crisis prevention and management and/or limitation of armed conflicts, and humanitarian measures.

272. In this context, the formation of a Franco-German armed force open to all the countries of the Union, as decided at La Rochelle on 22 May 1992, will help create a stronger European identity in the framework of Atlantic solidarity. This armed force is expected to be incorporated in due course into the WEU collective defence system when the WEU becomes an integral part of the European Union.

273. The EPP attaches great importance to the continuation of the CSCE process. It would like to see its institutions and its role develop within an expanded network of relations with Europe as a whole.

For a Europe open towards the world

274. Although the world situation (crises, conflicts, etc.) directly affects Europe, Europe itself does not always play the full part it should in world politics.

275. The European Community, the largest commercial power, will contribute to the development and regulation of international trade in goods and services on a basis of reciprocity, while at the same time ensuring that its legitimate interests are defended.

276. While continuing to seek full integration and develop their cooperation with the other nations of Europe in the areas described above, the EC states will try to maintain special ties between the two sides of the Atlantic.

277. The Member States of the European Community will attach particular importance to maintaining close ties with North America and moving towards a type of cooperation based on equal partnership. These ties are justified by their shared secular values, which have led to the development of parliamentary democracy and the universal establishment of human rights. Moreover, this kind of transatlantic understanding serves to help in the definition of a new peaceful world order based on greater respect for the rights of each and every individual.

278. In its relations with the rest of the world, the European Community will encourage and promote the conclusion of regional agreements and security, economic, social and cultural cooperation agreements, wherever

such agreements prove necessary to the maintenance or search for peace, the promotion of disarmament and justice and the economic development and improved well-being of the various peoples, on the basis of respect for their fundamental rights, even in regions where the European Community is not directly involved, such as Central Asia and the Far East, the Pacific, the Indian Ocean and Latin America.

279. While respecting the sovereignty of each nation, the European Community will promote human rights throughout the world. It will also give help to all those developing countries that are fighting to promote the cause of justice, defence of freedom and greater solidarity.

280. Conscious that peace is only founded on justice and that the new name for peace is development, the European Community will step up and improve its measures to promote and support economic and social progress in the Third World, particularly among those countries with which it has already concluded important cooperation agreements. Its development aid policy will extend and coordinate the Member States' financial and technological commitments.

281. The European Community must defend and promote in the United Nations the right and duty to intervene on a humanitarian basis wherever it finds that peace or human rights are being violated constantly, massively and brutally. The EPP will work to ensure that the United Nations is not only recognized as having legal powers but that it is also entitled to the logistical (including financial) support needed for the prevention of conflicts and wars, the effective application of sanctions, a permanent intervention mechanism and legal proceedings against individuals responsible for massive, prolonged human rights violations. The Community must act as a genuine regional organization for the settlement of disputes, as provided for in the United Nations Charter.

282. Lastly, the European Community will devote itself to setting up where they do not exist, and strengthening where they do, the institutions that prepare, coordinate, reinforce and regulate international and supranational cooperation, with a view to creating a universally recognized legal system. This new world system must be given real powers to deal with private or public contraventions. But it must always be based on respect

for human rights, which constitute the universal point of reference and governing principle of all political, economic, social and cultural action.

For a People's Europe

283. Given that the starting point of European political integration is our common image of man, the Constitution of the Union will have to supplement the provisions of the Maastricht Treaty relating to citizenship. The Constitution will have to define clearly the rights and duties of citizens, men and women, and ensure greater legal protection for fundamental rights and civil rights by enabling individual cases to be referred to the Court of Justice.

284. The Community (or the Union) requires the active participation of European citizens. Europe is not - and cannot be - the affair of governments alone. It is the citizens themselves who must think and make Europe.

285. In this context, the EPP particularly welcomes the recognition - incorporated in the Maastricht Treaty at its request - of the irreplaceable role played by the European parties: 'Political parties at European level are important as a factor for integration within the Union. They contribute to forming a European awareness and to expressing the political will of the citizens of the Union.'

286. The EPP seeks the intensive participation of the citizens of Europe and the creation and development of democratic structures. It firmly supports the commitment of many citizens to democratization and participation in political responsibilities. The direct election of the European Parliament forms an important basis in that context. Thanks to the creation of European citizenship, every citizen of the European Union will be able to vote in local and European elections in his place of residence regardless of his nationality.

287. The EPP undertakes to work constantly to provide European citizens with information about the unification process. It also undertakes to defend their interests through its group in the European Parliament.

288. Just as political parties are vital to the achievement of European unification and the political and social development of a common Europe, so unions, associations and other institutions such as churches must be seen as very important to the achievement of this aim. We encourage any initiatives within and outside the EPP that are aimed at European integration and are committed to the rapprochement of the peoples of Europe.

289. To guarantee the acceptance and long-term success of the European Union, the EPP calls on the younger generation to play a large part in the process of integration. It therefore undertakes to make every effort to promote youth exchanges and the mobility of young people. It also encourages young people to take part in youth organizations and support the work carried out at European level by non-governmental youth organizations.

290. Beside the political parties and institutions bearing public responsibility, the media also have an important and vital role to play in forming public opinion. Their task of informing, educating and entertaining is linked to respect for the fundamental values of our free society. They therefore have a duty and responsibility towards the public.

291. The principle of democracy accords with the pluralism of our societies. It must be upheld on the basis of common values and principles. At the same time we must ensure the protection of minorities and divergent opinions.

292. Europe must be free from racism and xenophobia. Faced with the resurgence of such sentiments and the acts of violence which accompany them, all political, social, economic and cultural groups must be aware of the serious challenge they represent to our society and must fight hard to overcome this form of intolerance.

293. Europe will not become more democratic until its citizens have a part in the decision-making process. That requires transparent decisions. The European Union has a duty to its citizens to provide comprehensible information and make it easier for them to read Community law, particularly by speeding up the process of its codification. Internal security and measures to combat crime.

294. One of the main tasks of the authorities is to combat crime vigorously. When people no longer feel safe, they have less confidence in the authorities. Crime prevention policy must be intensified in order to deal with violent attacks and offences, crimes against the environment, fraud and organized crime.

295. The EPP hopes that the European Union will be in the forefront of this activity, aware of the need to protect the foundations of our culture and our freedom for future generations. It therefore calls for the Union institutions to pursue a resolute policy in this respect. These institutions must be given appropriate powers of inspection and sanction, and there must be systematic cooperation between Member States' police forces and administrations.

- External frontier controls must be improved and carried out within a democratically-controlled legal framework.
- The system of tax and banking regulations within the territory of the Union must be such as to prevent criminals from evading their tax obligations or other financial control instruments.
- The policy on legal proceedings and police investigations must be very closely coordinated and a special organization to fight crime must be set up within the Union ('Europol'). It will thereby be possible to combat international criminal organizations, particularly those in volved in drug trafficking, more effectively.

296. A common immigration and asylum policy is a precondition for the success of such measures.

Ethics and technology

297. New technological developments represent new challenges to man in his attitude to the created world. Two of the main problems of our times are the way in which Europeans use technology and what rules the authorities should lay down in this field. Technological progress has brought great benefits, but as things stand, the relationship between ethics and technology needs to be reviewed, as regards:

- for instance, the protection of privacy and the attitude towards the possibilities and limits of innovation;
- whether to invest in research in Europe and the world on the basis of consumer supply and demand or other social or public interest requirements.

298. On the question of the ethical problems raised by technological progress, the EPP's position is based on the following considerations:

- A code of ethics must be drawn up in the field of genetic manipulation, the use of embryos for genetic engineering research and animal experimentation.
- Any trading in manipulated human material must be prohibited.
- The health care system must ensure that every individual receives the necessary care and treatment.

The cultural dimension

299. The EPP affirms that protection of the freedom of religion, social initiative and ideological pluralism must form the basis of European cultural policy. This policy must take a positive approach to the variety of intellectual and spiritual traditions that, taken together, and harmoniously linked by various exchanges of ideas and mutual initiatives, make up European culture. The European Union must treat the different intellectual and spiritual forces (churches, charitable organizations, etc.) as genuine interlocutors. While respecting the competence of the Member States and/or the regions in cultural matters, European cultural policy must encourage:

1 - cultural and artistic events on a European scale;

2 - awareness of European popular traditions;
3 - initiatives reflecting the sense of European cultural identity which binds together the various cultures;
4 - the European dimension of education and research and the promotion of the great traditional values of European culture;
5 - freedom of information and freedom of opinion as a crucial foundation for a free society which advocates pluralism and the dependence of the media and the maintenance of a non-commercial sector.

EPILOGUE

As Christian Democrats, members of the European People's Party, we want to make our contribution to building a Europe and a world in which the old hatreds and new resentments dividing people make way for cooperation and efforts to work together.

We ask every man and every woman - be they Christians or not, be they believers or not - to join us in order to build together, for our children and for ourselves, a better world based on greater justice, greater solidarity and greater democracy.

From the very start of the process of European integration after the Second World War, the Christian Democrat founding fathers of the European Communities focused on the fundamental human and social dimension of their vision of the future of the peoples of Europe. Forty years later, we can see that their vision has borne fruit on an impressive scale: European unification and the European Community have been salient factors in the history of the second half of this century.

European integration is based on principles which are an essential part of Christian Democratic thinking and are now widely accepted: subsidiarity as a guiding principle in social and political organization, the decentralization of powers, a social market economy, a respect for spiritual and ethical values, an opening up to the rest of the world and a respect for the created world.

We call on all citizens of Europe to work for the development of those principles within the European Union. We Christian Democrats, members of the European People's Party, wish to continue the task of building the European Union on the basis of those principles and objectives.

'A Union of Values', Basic Document Adopted by the Fourteenth EPP Congress in Berlin on 11–13 January 2001

A UNION OF VALUES

- final text agreed at XIV Congress

PREFACE

001. The last decade of the 20th century saw, at least in Europe, the triumph of liberty, democracy and the rule of law. The 21st century poses new challenges to our values. Globalization, the new economy, the Information Society and new technologies demand new answers. The EPP will respond on the basis of our traditional values. These have to be reaffirmed, rethought, and modernised in order to make them applicable in the 21st century. Pragmatism, efficiency, or some undefined 'third way' do not address people's real concerns. The European model is based on values, culture, and history. That is where the answers to the new questions come from. That is the starting point for us as Christian Democrats, moderates, and centrists, members of the EPP. And that is the European People's Party's new vision of a Union of Values. The 21st century offers Europe the chance to build a European Union worthy of the name - a Europe that is whole, free, and prosperous. The human person must and will be at the very centre of our politics.

002. The immediate challenges are to enlarge the Union, to include the new democracies in a Europe in which values are the driving force of progress and integration.

003. Full employment remains our main goal, because it is the best guarantee of social cohesion. Sustainable welfare systems must be built to ensure that a just and equitable society is also the birthright of future generations.

004. The explosive success of information technology creates another opportunity - a Society of Knowledge, in which education and a new spirit of enterprise are central.

005. The challenge is to safeguard what is truly, uniquely valuable in our civilisation as we embrace a radically different world, one in which many facets of people's lives are changing.

006. Coming from many different cultures and traditions, secular and confessional, the EPP's 39 member parties have remained united by certain core values: freedom and responsibility, dignity of the human person, solidarity, subsidiarity, justice, the rule of law and democracy as they were defined in the Athens Basic Programme.

007. The European People's Party, rich in its universal spirit, represents the strongest hope of finding new ways which are in keeping with the tradition of Europe's classic, humanistic and Christian roots. The EPP is more than ready to innovate because we are sure of our own identity. The process of transforming the European People's Party from a union of national people's parties into a genuinely European Party, is in the process of being completed.

008. The founding fathers of the Union - Jean Monnet, Robert Schuman, Alcide de Gasperi and Konrad Adenauer - were also the founders of the party. The EPP is the great European party, the only one capable of tackling the huge tasks ahead. Our political approach politics is based on dialogue and consensus. Now a new generation of political architects is needed, to realise our vision of Europe, and our conception of what it means to be a European.

009. That is the purpose of this document: to outline the party's vision of the future of Europe and the World and our response to it as a union of political parties - fighting for a Union of Values - heading towards a real Union of citizens.

I People at the Centre of the Union

101. Every person counts. The future development of the Union must have at its core the freedom and dignity of the human person. This is the only base on which we can build popular support for the process of integration. We must create a Union which protects and promotes common European values; and in which fundamental rights are respected and no form of discrimination is accepted.

102. Families where fathers and mothers take responsibility for their children are the foundation of our society. We recognise the existence of other forms of life communities and their needs. In order to give priority to the needs of children, and respect the equality of men and women, legislative and economic measures must be developed which enable parents to strike an appropriate balance between family and work commitments. The Party supports all measures to improve the position of children.

103. Inclusion. The party rejects all forms of racism, xenophobia, or social exclusion; considers that, on grounds of social justice and solidarity, it is the duty of the State to meet basic needs where the individual is unable to do so; and endorses the Charter of Fundamental Rights, which sets out both the rights and responsibilities of Union citizenship. Rejecting social exclusion the party will support the smaller and natural units in the civil society, from the individual and the family to the associations of civil society.

104. Equal opportunity is fundamental. Consequently all citizens must have access to the basic services of healthcare, old-age pension provision, housing, and education.

105. The Union must be open to those seeking asylum from persecution. Common rules on asylum policy in the European Union, based on the Geneva Convention, with a fair distribution of asylum seekers and refugees among EU Member States are of decisive importance for the EPP. On the other hand adequate measures must be taken to deal with the increasing abuses which cannot be tolerated within the Union. A transparent, planned approach to immigration and integration, taking account of the Member States' capacity to receive and integrate immigrants, is needed for those who come to live and work in the Union. The Member States should

elaborate an active policy to integrate immigrants who have entered the state legally. While accepting established European values such as the rule of law, equality of men and women, Judeo-Christian values and the values of secular humanism as a step towards integration in European society, people of other religious backgrounds may organize themselves on the basis of their own background. A controlled immigration policy must also embrace political and economic cooperation with the countries of origin, the fight against illegal immigration and an effective mechanism for surveillance of external borders. Europe is a pluralist culture in function of the unity of its values.

106. The population of Europe is living longer because of medical advances and falling birth rates, so that social security needs to be adapted to these changes. Clear and resolute support of the family, stimulating political and economic support for childcare and children's education would help to rejuvenate Europe and to turn around the current drop in birth rates. We believe that extending one's working life should be offered as a possibility in future. Ageism is a form of discrimination which is as unacceptable as other forms of discrimination. We have to encourage private initiative and the non-commercial sector.

107. A civilized Europe means a Europe based on the rule of law. Free movement must not mean that the Union, especially in the cities, becomes a haven for criminals. This means decisively reinforcing judicial cooperation and direct contacts between police forces of current and future Member States and, where necessary, European and international law. Organised crime and terrorism must be efficiently and jointly fought because they represent a serious threat to the peoples of a united Europe. We demand more rigour in the struggle against the most dangerous crimes and in particular such as the exploitation of prostitution and paedophilia.

108. It also means a Europe that enjoys quality of life. Policy must be directed to improving and preserving both the urban and the rural environment.

109. The political challenge ahead is to build an enlarged European Union based on common principles and values, a Union which puts the concerns of people at the centre of its work. A Union based on enterprise and social

justice and care for the environment will enable us to fulfill that dream which is now, for the first time, within our grasp.

II Europe's New Economy in the World

201. From the euro to full employment

202. Full employment and well functioning labour markets are the objective of the European Union which can be attained if people are given the opportunity to adjust to the structural changes of our technology and information-driven age. For the EPP, employment is more than just ensuring material existence. It is also a decisive factor of self-achievement and it opens up the opportunity of participating in the construction of society.

The Single European Currency is an important factor in a period of economic dynamism driven by global competition, technological advance, and ever-wider acceptance of the market economy. The euro has delivered the following key advantages:

203. Elimination of exchange-rate fluctuations and removal of competitive devaluations between Member States. Even before circulating as cash, the euro has improved the availability of capital, reduced costs, enhanced competition, and increased productivity. It has also softened the impact of external financial crises.

204. The independence of the ECB is vital to the success of the EURO and any attempts to subdue it should be strongly rejected. Its independence and the clear strategy in favour of price stability is the one single biggest improvement of economic policy making of the last century. But management of monetary policy should become more transparent and have more coherent external representation. This will strengthen the euro's standing as a global currency. There is a need for prudential control of the banking system on an EU level. The Central Bank´s neutral money policy must be balanced by a more coordinated economic policy on the part of Member States. But this must not lead to a single economic policy.

205. Balanced budgets as an obligation of the Pact of Stability and Growth need to be achieved by lowering expenditure rather than increased taxes. Budget surpluses should primarily be used to reduce debt and taxation.

The favourable economic trend should be exploited to create margins of manoeuvre, for instance to deal with the growing problem of an ageing European population.

206. The Euro is much more than a monetary project. It is also to a large extent a political project. The Euro is a policy of peace for the 21st century by monetary means. The credibility and the confidence which the euro should enjoy as a major international reserve currency will also depend on the EU's capacity to better coordinate its economic policies, which is an essential condition for convergence and the joint achievement by the Member States, acting jointly, of price stability and growth objectives.

207. Promoting competitiveness

208. In the framework of a social market economy, we support increased liberalisation of the European economy in order to complete the Single Market. In the particular cases of telecommunications, electricity, gas, and transport, the advantages for employment and for the consumer have been proven, and contribute to the promotion of the Information Society within an environmental and social framework.

209. The promotion of the European economy presupposes the improvement and development of infrastructure for all modes of transport, and in particular the extension of the trans-European transport network to the candidate states of Central and Eastern Europe.

210. Subsidies are not the way to create new jobs. As a general rule, resources must not be taken from growing areas of the economy to preserve old jobs, but must be concentrated on providing opportunities for training and specialization. Sustainability demands that we invest our credits in the future and alleviate the social impact of a sector whose time is past rather than maintaining no-hope-sectors. Any legislation discouraging private initiative must be avoided.

The state's obligation to provide some public services for people must not be unduly restricted by European rules. The local, regional and national communities must be enabled to decide about the services they provide for its citizens. This is an important part of the European model and forms part of the political scope of local and regional authorities.

Hence, cutting subsidies and real liberalisation of, for example, the postal service, the railway systems, banking, housing, and waste management sectors will bring the public welfare witnessed in the telecom and electricity markets.

211. Spirit of enterprise

212. Innovation. Freedom is also freedom for enterprise. The human person is at the centre of our view of the new economy. We must combat materialistic thinking, reducing the human person to a consumer or a client. Personal initiative is the most important element in generating work and wealth for everybody. Europe's best chance of a permanently high level of exports and low unemployment is to create new, high value-added, products or services. Europe cannot, over the long term, compete with low-wage countries manufacturing products which do not incorporate advanced technology. The EPP is against subsidizing loss-making enterprises on such pretexts as maintaining employment. This distorts the market and diverts resource from research, training, and investment in competitive enterprises.

213. More competition decreases prices. The cost of capital can be reduced by monetary stability, lower interest rates, and better organisation of the capital-markets. A general European legal framework for companies will also reduce the cost of investment. Opening national markets to competition will reduce e.g.: energy and capital costs. Increased rationalization of public services can reduce costs and improve services and customer orientation. Better access to risk capital will be essential to support growth sectors and job creation. Finalising the internal market for financial services and foster the use of venture capital and stock options for employees thereby is one of the prerequisites of improving Europe's competitiveness. As things stand, taxation and other levies are much higher in Europe than anywhere else in the world. The EPP intends to reduce them.

214. Encouraging use of the Internet. General telecoms policy should seek to support the liberalization and privatization of markets, so that increased competition can bring down prices and encourage the growth of the IT economy.

215. The right framework for enterprise. The high cost of setting up an enterprise in the EU is a massive disadvantage for Europe and a key reason for the eye - popping difference in employment rates and growth with the United States. Direct production costs are only one factor affecting competitiveness. Others are: the quality of education and training, the efficiency of public administration, transport and communications, research, people's propensity to save, wider share-holding, and - something which cannot be legislated for - the solidity of the social consensus. All EU governments can learn from successes beyond their own borders. Therefore there should be EU bench marking studies on all these factors measuring not the state expenses but the effectiveness of each euro spent.

216. Policies aimed at creating durable new jobs. Non-labour costs are one of the major causes of unemployment, especially amongst unskilled workers. Such burdens need to be reduced by cutting social security costs through improved management of social security expenditure and by reducing people's tendency to think in terms of what entitlements they can claim. Wage increases must be clearly related to increased productivity.

217. Flexibility, employability

218. Labour flexibility is essential in a fast-moving and unpredictable global economy if redundancies and / or higher production costs are to be avoided. This is especially true in the new, fast-growing, but volatile, IT sector. The EPP therefore supports initiatives and derogations which free SMEs from excessively burdensome administrative regulation, and encourage them to recruit staff as SMEs are the one most important single source of job growth in Europe. Part-time work while safeguarding social protection equal to full-time work should also be encouraged for men and women, both as an element of flexible labour and as a means of reconciling work with family life. The EPP supports the idea of a European basic pillar of labour rights. The EPP supports the initiatives taken by the EU social partners to find mutually acceptable measures to deregulate the labour market.

219. Employability. Most of the durable "real" jobs cannot be created by the state. While claiming to do the opposite, job creation by the state or agreeing to such proposals as reductions in working hours without loss of

income can deteriorate overall employment. Policy should rather concentrate on training and retraining, as well as the promotion of mobility, and adaptation. Policy must not interfere in the wage negotiations. The EPP therefore strongly supports the independence of the social partners. However the EPP clearly recognises the trend towards collective bargaining on the regional and company level. In the shorter term, linking unemployment benefits to individual efforts to find work or acquire marketable skills, and encouraging young people to set up their own businesses, have all demonstrably produced results. Labour market rules must be reformed in order to support new working patterns. Such labour regulations that are tailored for an outgoing industrial society will be an obstacle to efforts to make Europe more competitive.

220. Education and training

221. In a knowledge-based society the opportunity of education is the key to progress and equality and sustainability. For the EPP it is not only a way of becoming more competitive. Education and training need to be widened as a concept and as a policy area, and to be understood as a lifelong process and a personal investment. The transmission of Europe's humanistic heritage to new generations must be an essential task of the educational system. This can be encouraged by special education savings accounts.

222. Europe should take a lead as the new economy develops. In doing so, the EPP realises that major efforts have to be made to change the way knowledge is being taught and spread. The demands both of academia and industry must be met. A close co-operation with the labour market is necessary in order to match education and needs.

223. Universities and other scientific and technological institutions must be given better possibilities, both for education and for research. Since education policy remains the responsibility of the nation states and regions, the European Union shall focus its policies on the extension of its exchange andresearch programmes. These programmes are essential in providing opportunities for students and scholars to mutually exchange information, which is decisive for high-quality research and high-quality learning and teaching. The European Union should take initiatives aimed at getting member states to agree on a division of work. The most ad-

vanced resources have to be concentrated in order to create the scientific environments needed.

Systems for the mutual recognition of degrees such as ECTS shall be extended to all sectors of education. In order to achieve the success of its mobility-schemes, the European Commission shall implement a policy to remove remaining bureaucratic obstacles enforced by the Member States.

224. Information technology is a remarkable tool not just of commerce but in helping schools to put more emphasis on teaching students to learn for themselves, and to draw their own conclusions. Every school should be connected to the Internet, and every student should ideally be given personal access. Europe is lagging behind in this area and it is deplorable that the continent of culture and history is lagging behind in the decisive area of e-content which means what is actually on the web. A concentrated effort by political, business, cultural and academic circles is needed to successfully face this challenge.

225. Schools must give each student a solid base for continuous learning. School education of the highest quality will be more important than before. Schools must also teach students to learn. This means that more emphasis must be put on understanding, problem solving and the evaluation of information. The family must be encouraged to work as a vehicle of education and assistance to the individual in accordance with the principle of subsidiarity.

226. Initiative and independence are a key to success in modern society. This principle should also apply to the education system. The state should encourage a variety of both schools and of teaching methods. Students must be given the right to choose their school based on their own interests. A European-wide system of school choice should be encouraged as well as a system which enables university students at European level to select the university of their preference.

227. Education and learning must be given the highest priority when policies for the new economy are being formulated. Despite the fact that learning is not a first hand EU-responsibility, exchanging information and experiences within the EU could be useful for developing constructive

national strategies. Acting as such a partner, the EU could assist both in making use of the variety of traditions and experiences of different member states and to combine forces where this is necessary to make Europe as a whole competitive. The EPP considers that the establishment of a European educational area contributes to the promotion of a Knowledge-based Europe, encouraging cooperation in the field of lifelong learning.

228. The European Social Model

229. Europe will have no meaning unless it is both an economic and social Europe. The social market economy links the market mechanisms of supply and demand with the obligation to respect the dignity of every human being. The values of the European social model (performance and social justice, competition and solidarity, personal responsibility and social security) remain relevant in conditions of globalized markets and rapid industrial change.

230. The challenge is to combine powerful new market forces with humanity - economic dynamism with social responsibility - in the way we organise our social security systems. The EPP believes this is possible and essential to more favourable combine conditions for entrepreneurship and economic dynamism with social responsibility. We will therefore seek to reform and modernise European social security systems. Social peace must be based on social partnership, which covers participation and joint responsibility of the social partners as well as of the government. The EPP supports negotiations between social partners on framework agreements at European Union level. These agreements should be made legally binding for the Union as a whole.

The main responsibility for social security lies with the member states. The EPP however, welcomes the fact that member states exchange their experiences at European level and diminish - wherever possible - existing qualitative differences between the systems of social security. Following the principle of subsidiarity, the Union must take those responsibilities in this field which cannot be dealt with by the member state alone. On the international level the European Union should actively promote a strengthening of social standards. Within the framework of the World Trade Organisation internationally binding rules on competition and codes of conduct in the fields of social, ecological and human rights standards must be progressively developed.

231. Europeans live longer an healthier lives than ever before. This is a great achievement in quality of life. But the demographic situation also poses new challenges as the working population decreases and the number of retired persons increases. This needs to be provided for in financial and fiscal planning. The EPP is committed to defend the principle of justice between generations. The society of the future will more than ever need human potential, that of women as well as men.

232. The other special concern is also more familiar: the poor and the marginalised. Generally speaking our modern society is very prosperous. Nevertheless the division of society between rich and poor must be overcome. The EPP believes that individuals have a duty to take responsibility for themselves, and to earn a living if they can. We are equally committed to the belief that a just society helps the helpless.

233. While acknowledging their shifting importance, the EPP accepts the legitimate and useful role of the social partnership of trade unions and employers' organisations in creating durable employment. The EPP's attitude to work goes beyond mere jobs. Work is a means of individual self-realisation and of playing a part in society. The European Union, as much as every Member State, should be guided by this perspective, and aim for the objective of full employment.

The inseparable connection between economic and social order becomes particularly clear in the endeavour to bring as many people as possible into work. Therefore the EPP is convinced that working conditions which offend human dignity must be prevented in the same way as the temptation to regard work only as a matter of price. At the same time it means the obligation of the European Union as a whole as well as of every Member State to fight unemployment and strive actively for more employment. In this context the EPP strongly emphasises the procedure of European employment guidelines implemented by Christian Democrats, which combine common objectives - the obligation to take concrete measures as well as the exchange of experiences.

Greening the Economy

Though industrial expansion has generated unprecedented economic gains there are many related environmental costs that have yet to be paid. The challenge is to bring harmony between economic growth, social development, and environmental concerns. Hence, the need for public interventions to integrate environmental costs.

III. Challenges of the Information Age

301. The Society of Knowledge

302. Information technology increases the potential of the person. The EPP supports the freedoms which are embodied in such potentials. Equality demands that everyone be given the opportunity to profit from the new possibilities. Information technology results in greater productivity. New jobs will be created wherever there is systematic training and education in the new technology.

303. E-commerce in particular is developing exponentially as an efficient market mechanism, and enhancing sound economic management. All barriers to e-commerce within the EU should be eliminated.

304. The success of the new high-tech economy depends on low telecommunications costs. High costs in Europe are a result of inadequate competition, on which the EU should intervene. The target should be to achieve open cost-free access to the Net as soon as possible.

305. Joint European action needs to be taken on security of contracts, confidentiality, and signature authentication; measures are also needed to prevent tax evasion or fraud, in particular regarding transactions based on non-localised services; and also against use of new technology for money-laundering, illegal trafficking of humans, drugs, and prostitution.

306. This implies extending EU powers when negotiating international agreements in the services area. Given the international, or even global, character of e-commerce, purely national legislation is inadequate. Member States must present a united front at international level.

307. But the Information Age also poses new social dangers. The growing "virtual" character of business, and new types of employment such as teleworking, pose risks such as covert electronic intrusion into consumers' and employees' private lives, and manipulative advertising, especially that one aimed at children. Where appropriate under the subsidiarity principle, EU-wide regulations should be drawn up on these matters. The gathering and the use of information owned by or related to individuals should always respect the concept of privacy, as understood in other parts of communication.

308. Technically, the European telecommunications market, until 15 years ago divided into national markets dominated by state monopolies, has largely been deregulated and integrated thanks to EU initiatives. These have also ensured that the GSM standard has become the dominant standard; for the same reason the UMTS standard will soon achieve a leading position.

309. Opportunities for everyone. The information and knowledge society brings about many new opportunities. It is important to make these opportunities available to all citizens. Therefore member states and the EU should redouble their efforts to encourage dissemination of new information technology. Europe should aid developing countries to participate in the world of Information Society.

310. The "New Democracy"

311. The information revolution will have important consequences for the way our democracies work. New and more direct forms of relationship between citizens and politics will emerge, bringing public administration, parties and the political actors closer to European citizens. The new technologies will bring new ways of conducting politics, from collecting funds to following political rallies through the Net from any part of the world. This fact will change the traditional ways of running political parties, especially during election campaigns.

312. The EPP supports the implementation of electronic voting as an option in the next European elections, and progressively at a national, regional, and local level.

313. The EPP will strive to use the new technology to give citizens full and updated information about its political work. It will also strive to simplify the means of communication between voters and our politicians at all levels. We believe that technological innovation opens up new possibilities for an intensified and meaningful democratic dialogue in society.

314. Administration on line. Administration at all levels should be able to relate with the citizens as quickly as possible. Paying taxes, applying for a position in the public sector, or resolving "red tape" questions, should be possible for everyone via the Internet. This will also ensure greater transparency in politics.

315. Bio-ethics poses a special challenge to the EPP

316. In the field of bio-ethics, the EPP recognises the exciting progress of science and technology, which contributes considerably to health and welfare. The EPP follows the principles of the protection and promotion of human dignity and, consequently, respect for the right to life and the uniqueness of every human being from the moment of conception to death; the special responsibility of parents and families; human beings' fundamental equality; the promotion of health, and the freedom of research and scientific investigation. The dignity of the human person implies that science is subservient to the human person and the human person is not subservient to science. Distinctions between "human person" and "person", or between the embryo as "a potential human being", cannot be made without introducing an unacceptable kind of discrimination.

317. The free of movement of people, goods, capital, and services in the European Union affects the possibilities of national governments to guarantee the dignity of human life in the field of bio-technology. Therefore, there is reason for involvement of the European Union in setting common legal standards concerning the protection of the dignity of human life and the responsibility of biomedical research.

318. The EPP supports a political course which seeks to develop guidelines and rules that promote bio-medical research in favour of guaranteeing human dignity, health and welfare. New knowledge of the structure of living matter and the possibilities opened up by the techniques for modifying such matter must be used to benefit human health (provided that human

dignity and freedom are always safeguarded) and must not serve as a means to an end which is alien to human life. EPP policy is an alternative to uncritical support for technological developments, in which the human embryo is seen as an instrument, an alternative to the politics of endorsing unlimited individual self-realisation, and an alternative to the policy of closing doors to any new development. This attitude runs the risk of being overtaken by reality, and as a result influence on research and the ethical debate could be minimal.

319. All individuals must be protected at every stage of life, from conception to death, particularly if they are weak, handicapped, or powerless. Human life, in whatever form, whatever its appearance or capacity, is dignified in itself. Any deliberate form of eugenics must be banned.

In the field of biotechnology, without using cells from living embryos experiments on human embryos should only be permitted if they are designed to protect the life and health of the specific embryo which is the subject of the experiment, as is the case with all other human individuals. Very strict world-wide rules should be adopted. The EPP is in favour of enhanced and joint efforts to promote preconception research techniques which will make research on embryos unnecessary, and of encouraging scientific research in the field of artificial insemination, which avoids the serious problem of overproduction of embryos. Commercial exploitation of human embryos must be forbidden in all cases.

320. The EPP refuses to consider abortion as "a solution" to problems created by unwanted pregnancies. The EPP promotes programmes and initiatives to help parents and families to accommodate every child, in particular when difficult or unwanted pregnancies occur. No pressure should be exerted on parents who decide to accept a child with a handicap. Modern pre-natal diagnostics must be used only after the parents concerned have had thorough consultations. They are entitled to professional, humane, and life-protecting guidance that supports them in making the choice which is best under the given circumstances.

321. Research in the fields of bio-medicine and bio-technology promoted by the European Union, and the resources provided by the Commission for

bio-medical research, should be granted in accordance with the ethical principles set out above.

322. The EPP supports strict controls on food safety. Europeans have a right to be confident that the food they buy is safe. But the growing market for "bio" food has a political and economic message which goes further than understandable scares about BSE and dioxin. People are increasingly interested in what they eat, and prepared to pay to eat healthily. Food marked as biologically produced must proof that this claim is absolutely valid, preferably by established EU standards controlled at national level. Europeans have a right to accurate information on product ingredients, particularly GMOs. Therefore, correct labelling of foodstuffs which allows consumer choice must be the priority.

Consumers need to be better educated and given current information on the scientific aims, GMOs and their effects on food quality, health and the environment.

323. The EPP therefore believes that new thinking must be supported, by promoting the potential of traditional agriculture and particularly by realising the value of typical traditional quality products. The EPP supports encouraging young farmers and diversifying production to satisfy consumer demand. Food safety, quality requirements and reliable information for consumers should all be taken into consideration, irrespective of the type of agricultural production used. In Central and Eastern Europe, an unthinking policy of modernisation will mostly serve to drive people from the land, and add to the glut of food in Europe.

324. New rural areas

325. Information Technology offers enormous opportunities to rural areas. As information is available everywhere, the role of the centres fades. The EPP firmly supports a pro-active policy for rural areas to make use of these opportunities. Rural depopulation, and the general malaise on the land affects everyone, including urban dwellers, disrupting their quality of life and even their sense of identity. The EPP is against a subsidized theme-park policy to the land, and favours policies which encourage a genuine rural economy. This will necessarily include more light industry in farming areas.

326. Traditional agriculture in order both to improve the quality of agricultural products for human and animal feed, to protect consumers from unsafe innovations and to promote consumers' consciousness remains a very important branch of the economy. Agriculture is not an ordinary branch of industry. Agriculture is more than just producing goods. Agriculture is culture in rural areas. Therefore we support and defend the concept of multi functional agriculture. There is a difference in farm sizes, social and environmental regulations etc. The EU farmers must be prepared to cope with world markets in a way which enables them to cope with this challenge.

IV European Identity in the 21st Century

400. The choice for the European Union is the result of history, and a response to the challenges of the future. Through the European Union, we intend to :

- develop a system which - in the era of globalization - combines a free economy with a society based on solidarity;

- consolidate and stabilise peace throughout the continent of Europe;

- make a contribution to a better world.

401. Over one and a half millennia, Europe's people have evolved into distinct and self-confident nations. Despite their national differences, they have preserved their shared cultural heritage - rooted in Hebrew prophecy, Greek philosophy and Roman law, as they have been harmonized by the Christian message and Judeo-Christian values. It has remained intact throughout each phase of cultural development: the Middle Ages, the Renaissance, and the Enlightenment. With the nation state, we have equipped ourselves with a form of political organization, and a way of life, which has become a model for the whole world.

402. But nation states on their own are less and less capable of safeguarding welfare, internal security, and peace. As a legal community, the European Union safeguards peace, enhances the interdependence of nations, and creates the conditions for their welfare and prosperity. The challenge for the future is to create a new form of political organisation.

403. European Union is the Europeans' contribution to the challenge of globalization. The European project is not to destroy the nation states of Europe, but to avoid nationalism and restore to them, through union - the real capacity to fulfil their central responsibilities. Europe is the prerequisite for the solution of the fundamental problems our nations face. As the essential means to this end, Europe is also an end in itself. In order to clarify the relationship between the European Union, its Member States and its citizens, the European Union needs a constitution which is treaty-based and compatible with the values of the rule of law, democracy, subsidiarity, and accountability.

404. The EPP advocates a governance which is capable of acting to realise our values. It follows from globalization and the Europeanisation of life that a European level of state-action has to be sustained not only to create a free trade area, but to set up a framework which generates freedom, equal opportunities, solidarity, and sustainability in so far as the nation state alone is not able to secure that anymore.

405. Europe must have the capacity to act if it is to safeguard the nation as a way of life. This is the basic condition for its acceptance by our citizens.

406. But Europe must only take on those tasks which can be dealt with solely - or at least far more effectively - at European level. As far as possible, Europe should restrict itself to fundamental and/or framework decisions. In a Union of up to 30 member states EU must limit itself to the core European tasks. Increasing diversity and a mounting work load ask for streamlining activities if the essential tasks are to be fulfilled. A lean Europe is built on self-governance by local and regional authorities and the enduring power of the nation state.

407. Europe must be democratic. Its institutions must reflect, directly or indirectly, the will of its citizens. It must function in a way which citizens can understand, and its procedures must be transparent. Citizens must also have a real say in influencing the Union's policies through their European Political Parties. It is also important to ensure fundamental improvements in the openness of EU institutions.

408. All three elements - a capacity to act, a restriction of action to certain key tasks, and democratic procedures - form the bases of Europe's legiti-

macy and are an integral part of its identity. In all these respects, Europe epitomizes not the disintegration but the evolution of its constituent nations.

409. Europe must further develop its existing common cultural and political identity. Europe defines itself in equal measure through its common intellectual and spiritual heritage, its diverse national origins, and its will for a common future. European civil society must be reinforced. Civil initiatives for social purposes have to be encouraged. The ties between individuals have to be strengthened.

The EPP considers that the establishment of a European cultural area centred on respect for and promotion of cultural and linguistic diversity contributes to the ideal of the European Union, namely to unite the citizens and peoples of Europe in respect for their culture and their national and regional traditions.

410. Europe's constitutional form and structure

411. Europe's future is not merely an extension of the national origin. As Europe is a new form of political organisation, sole, ultimate, or universal responsibility will no longer exist at any level. Europe will not be a federation in the conventional sense, but a new form of federal system, of which economic and monetary union is the precursor. This unique form of political organisation will follow the objectives of federalism and the community method.

412. European federalism is based on solidarity between Member States but also on competition in order to establish best practices. The aim is to learn from each other and to establish Europe as a learning community. Economic and monetary union has centralized just one element - monetary policy - of one key area of public policy, the economy. The others, in contrast to a traditional federation, remain the responsibility of participating states.

413. Competition and solidarity are two mutually dependent elements of this federalism. The fundamental expression of solidarity in EMU is adherence to the rules agreed in the Maastricht Treaty which aim to modernise and revitalise European economies and societies. But within this

framework Member States are free to decide how to achieve this goal. Member States have to be given enough room for manoeuvre to adapt themselves to new challenges in respect of their specific problems, needs, and experiences. Harmonisation in the field of economic policy must ensure that a level playing field will come about for economic actors. Within this new federal system of economic and monetary union, the objectives of federalism - welfare, equality, and justice - are therefore not primarily the outcome of transfers and harmonisation as in classical federal states, but the result of a fair and rule-based competition between the Member States. The division of competences has to be clear, precise, transparent and dynamic over time. The citizen has to know who is responsible for what in Europe. Subsidiarity must be comprehensively applied.

414. The European People's Party advocates a thoroughgoing institutional reform which will ensure the proper functioning of the Union while taking account of the values of freedom, democracy, responsibility, equality, justice, solidarity, subsidiarity, the rule of law and accountability which are the values guiding its political action.

415. The European Union has need of a constitutional treaty to define the decision-making procedures among the European institutions and the allocation of powers between the European Union and the individual Member States, in accordance with the subsidiarity principle. This constitutional treaty must also include a Charter of Fundamental Rights. This constitutional treaty, which should be able to be adopted by the peoples of Europe in accordance with the most democratic procedures, should be drafted by a Convention similarly composed as the Convention for the Charter on Fundamental Rights.

416. The borders of the European Union

417. The sense of belonging together is nurtured by shared experience of the past - even in conflict and discord, the shared experience of the present, and the imagined experience of the future. This three-fold experience is interpreted in the light of common cultural values and convictions which extend far beyond a commitment to human rights and democracy. The more these values and convictions resemble each other, the greater the sense of a common identity.

418. The European Union should be an open community to European countries that would like

to join. The enlargement of the union represents an opportunity to safeguard freedom and democracy in those parts of Europe where dictatorship previously destroyed the lives of millions of people. An enlarged union is an important opportunity to secure peace in our part of the world.

This process gives rise to a certain fluidity of boundaries. Nevertheless, clear defined borders remain essential. The borders of the EU will be the result of a political process of shaping a self-confident European community whose members share a common sentiment of belonging and acting together. At the same time the Union has to develop a coherent strategy for the policy towards their neighbouring countries, which are bound to become important partners. In many cases relations based on the specific needs and interests seem to be much more appropriate than membership.

V A Europe Open to the World

501. The EU is a global economic player but very weak in other areas of foreign policy, notably security, intelligence, and sometimes even diplomacy. This imbalance creates a dichotomy in public attitudes to the EU and contains latent potential for conflict. The EPP recognizes that the European Union can only survive as a political project if it evolves into a Union with a dynamic Common Foreign, Security, and Defence policy. The EPP believes that politics have a general responsibility in establishing a framework regulating liberalisation and globalization to pass from growth to welfare for all. It is an absolute necessity if globalization is to be accepted by all countries and peoples. Cooperation must be promoted between the WTO and other international institutions, including the Bretton Woods institutions, ILO and other UN organisations, with a view to attaining a more consistent approach to trade, social, monetary, financial, ecological and political issues.

502. Europe must take responsibility for asserting its vital interests in global freedom, peace and prosperity. All European nations agree on this point, but cannot in practice play a major role on an individual basis. Only Europe - as a unit - can do so, and this is also what is expected by the

European public. Europe must match its economic power with the political, diplomatic, and military resources required to represent its interests and exercise its responsibilities. The EPP supports every effort to achieve this goal, and urgently calls for this process to be accelerated.

503. Europe's strength

504. Europe is a model for the rest of the world in legally regulating cooperation between free and self-confident nations, and in safeguarding peace to their mutual benefit as a response to supranational reality. Europe is the most advanced project of political integration, and the Europeans can be proud of their achievement.

505. Representing European interests in the world is one of Europe's prime tasks. Progressive globalization requires regulatory framework. Many international organisations such as the UN, the World Trade Organization, the International Labour Organisation, and international environmental projects, have already made a start on solving global problems with global polices. The European Union must represent European interests in such areas effectively, since Member States cannot achieve this on their own.

506. In the long term, a better world cannot be based on hegemony, a balance of power, or deterrence. Europe can - and must - make a major contribution to the urgent task of developing globally binding rules in areas ranging from security policy to the environment and the economy. The United Nations must therefore be strengthened through comprehensive organizational and structural reform. It can contribute much to the gradual process of replacing anarchy around the world with the rule of law, as it did in Europe itself. The EPP supports a fundamental reform of the UN and a single Security Council seat for the European Union.

507. Peace and democracy

508. Europe must learn to think and act on a long-term basis. A better world is above all a more peaceful world. Peace is based on freedom and respect for human dignity, and on democracy and justice. Promoting these values and helping to establish structures committed to social justice, democracy, and the market economy must be a key foreign policy goal, both of the EU and its Member States. The resources needed for this purpose

must be substantially increased. They are modest compared with the costs of the military and its deployment, and represent an investment in a shared and better future.

509. Solidarity and cooperation

510. The development policy of the European Union and its Member States is the expression of Europe's solidarity with other parts of the world. Europe must not only pursue an active policy of economic cooperation with the countries of Africa, Asia, and Latin America, but also provide assistance with democratization, the creation of a eco-social market economy, and the establishment of a functioning public administration to the states in these regions. In this context the charge of coordinating and integrating the development policies of the Member States must be assigned to the EU, in order to improve the efficacy of initiatives and to encourage a more rational use of investment. Moreover, Europe - the EU and its Member States - must be the advocate and champion of its development policy partners within the bodies of international organisations. The arms trade is one area where Europe has a particular responsibility towards the safeguarding of human rights internationally. Governments must ensure that profits made from the arms trade are not gained at the expense of human rights.

511. For its part, Europe will take its responsibility. The EPP strongly urges all Member States to meet the internationally agreed standard to devote 0,7% of their gross domestic product to development aid. The European Union should open its markets further to import products from developing countries. This is fundamental in helping poor nations to combat poverty.

512. Cancellation of Third World debt. The EPP is in favour of a conditional cancellation of Third World debt. That implies the drawing-up and implementation of specific development plans for each country; their acceptance should be a precondition of cancellation. Democratisation as well as good governance"] and enrolment in schools must be the main common pillars of these plans as the only ways of investing in the future. Investment in these countries should primarily be directed to the support of grass-root SME networks as they are the only solid basis of a sustainable growth per-

spective in developing countries. The establishment of the rule of law and the respect of the right of property and its protection is essential.

513. Joint European crisis management forces

514. EU foreign policy must be backed by military resources. The widespread use of force for non-defence purposes - for instance, to oppress minorities within a state, or to perpetrate other grave violations of human rights - must be met with sanctions by the international community. The EPP recognizes the concept of Human Security as a principle of at least equal weight to State sovereignty. However, as military intervention invariably can - at best - only create the preconditions for a political solution to the problems underlying the conflict, it must be subject to careful political as well as military scrutiny, and action must be taken in accordance with the principles of the United Nations Charter.

515. In a globalized world, the impact of wars and conflicts is more difficult to contain. Deterrence and conflict preventionare the real purposes of the armed forces.

The EPP calls on the Commission to submit budgetary proposals to the Member States so that investment expenditure in the field of new military communications and transport technology can be planned jointly, thereby reducing unit costs and increasing the operational capacity of the future European army.

516. Together, EU Member States have armed forces which numerically exceed those of the US. There should be greater and more consistent investment in European armed forces since these increasingly threaten to become second rate armies. Now that the Cold War is over, it is paradoxical that the countries which decided to combine their destinies within the European Union are finding it easier to reach an agreement on preserving peace outside the Union (the "Petersburg missions") than on the joint defence of the security of the Union itself.

517. The EU is a common security area; its members have common security policy interests but lack the capacity to act on an individual basis. Common European armed forces could be a smaller but more effective force. This is the way forward following the establishment of the Euro-

corps, and the decision to set up a 60,000-strong joint force for crisis-management.

518. Joint security analysis, joint long-term force planning, adequate funding, joint planning and procurement of equipment, a common defence industry, and a common arms exports policy, are crucial.

519. The security of Europe also requires the improvement of our co-operation in civil crisis-management, such as police and rescue operations. The EU therefore must be provided with a common police-force. Further, it is desirable that Europe improves its co-operation in crisis-prevention and democratic institution-building.

520. Europe and America

521. Democratic ideals have now taken root in the eastern part of Europe. Europe and America are jointly extending the Euro-Atlantic structures of the EU and NATO to this region in order to safeguard long-term peace, stability, and welfare there too. Both partners are also seeking to help Russia and Ukraine to find their place in Europe.

522. The key strategic project in shaping Europe as a whole is eastward enlargement of the EU. The institutional guarantee of peace and stability is the EU's greatest challenge and it requires American support. NATO has already admitted new Central European Members. NATO reform must progress towards an Alliance between America and a Europe capable of acting as a single European unit.

523. Anchoring Russia in Europe and promoting democracy and a social market economy are key interests for Europe and the USA alike. Europe requires this support, above all, in view of Russia's situation. Through the Partnership and Cooperation Agreement, and the NATO-Russia Founding Act, the EU and NATO have clearly signalled their willingness to enter into partnership with Russia. Partnership also entails being able to voice opposition.

With regard to the EU's largest neighbour, Russia, the EPP must seek to build on the Partnership and Cooperation Agreements (PCAs) with Russia and other former republics of the USSR.

The EPP will also seek to develop the EU Common Strategy on relations with Russia which will give further strength to the PCA in areas of political, economic and trade relations and lay the basis for cooperation in the social, financial, scientific, technological and cultural fields with a goal of incrementally liberalizing trade throughout the continent. This must be understood in the context that the EU is the major trading partner for Russia and most of the newly independent states, accounting for 40 to 50% of their global trade.

524. Sharing the real challenges to both Europe and the USA outside Europe is a natural implication of a European Common Foreign, Security, and Defence Policy.

525. Europe and Africa

Africa is a priority for Europe. The dramatic situation in many African countries - economic and social situation, conflicts and increase of refugees and internally displaced persons, declining health services and growing threat of HIV/AIDS - concern all European countries.

Therefore, we must continue the Africa-Europe dialogue and co-operation to integrate Africa in the world. It means :

• Integrate Africa into the global economy;

• Promote regional economic co-operation and integration;

• Develop a European development policy for sustainable development and poverty eradication;

• Promote human rights, democratic principles and institutions, good governance and the rule of law;

• Implement peace-building, conflicts prevention and resolution.

526. Europe and Latin America

The European Union should strengthen its relations with Latin America and give a positive and clear signal to the expectations of our partners in the region. Support for regional integration should be an important part of our policy.

527. Our view of the world

528. Europe must undertake its own joint evaluation of global security. The proliferation of modern military technology, especially weapons of mass destruction and delivery systems - also in unstable countries with aggressive regimes - domestic conflicts which spill across borders and trigger migration, and the links between unstable countries with aggressive regimes with terrorism and international organized crime, pose a threat to our security.

529. The EPP, however, tends to a more optimistic view. Without under-estimating the dangers, there are powerful positive trends in global devel-opment, and new opportunities for a better world.

530. To avoid new conflicts on a global scale, a comprehensive range of cooperation and assistance from the West, along with a policy of conflict prevention, regionally and globally binding security policy agreements, arms control and disarmament, are vital. Europe's experience of living through and overcoming this phase in its history means that Europe can and should make a significant contribution to this.

531. The overwhelming majority of nations and leaders are aware of the need to adapt to the modern world, and that by promoting economic devel-opment they not only enhance their countries' welfare and prosperity but also change the structure of their societies, and that this increases the desire and the need for democratic participation. They know that they need international cooperation and help to achieve these goals. Europe serves as a shining example of allowing nations to go their own way, and respects their desire for a global order based on diversity and multi centrism. Despite the ambiguities and setbacks in the global development of civiliza-tion, the European People's Party believes in the opportunity for a better world and is convinced of the need for an essential European contribution to this process. The real challenge for the EU's Common Foreign and Security Policy is to develop global policies in genuine partnership with the USA.

532. Geopolitics is an important instrument of political analysis. But it should not be the determining factor in international politics. Our view of the world is based in the human person and our values.

533. Environmental audit is needed. Europe is also vulnerable to environmental destruction caused by the industrial civilization. The answers can only be found when environmental aspects are integrated into the economy. New technologies can also be a strong ally in reducing pollution and environmental degradation.

The EU must continue to play a leading role in the fight against climate change and must also endeavour, to that end, to obtain the ratification and implementation by the USA of the Kyoto Protocol. We must do all we can to ensure that the EU adopts legislation on environmental responsibility which is based on the principle that the "polluter pays".

534. Europe and Islam

535. The Islamic world, from Morocco to Iran, is Europe's nearest and most important neighbour. Its rich culture, its political and economic potential, and its internal conflicts make this region, in the EPP's view, the greatest long-term challenge for European policy-making. This acquires an extra dimension through the growing presence of Islam within European societies. Cultural dialogue with Islam is therefore essential in safeguarding peace and tolerance within the European nations, and in promoting good-neighbourly relations with Islamic nations. More access to the European market and more assistance, above all in developing democratic and market economic structures, are further key elements in achieving this goal.

536. The European Union cannot neglect the need to play a greater political role in this region. This applies especially to the Middle East, where the proliferation of weapons of mass destruction will pose security problems for Europe even sooner than for America. Within the field of European defence, a new form of non-confrontational deterrence must be developed in conjunction with America. Within the field of European diplomacy, strategies for disarmament and arms control, as well as comprehensive security policy agreements, must be developed for the region. However, these ultimately depend on a solution to the conflicts underlying the build-up of arsenals within the region.

537. Europe's history and especially its geographic location together with its global responsibilities make it incumbent for i t to play a key role in the

Middle East Peace Process. This must extend beyond the provision of necessary financial resources, so that the Essen European Council's decision to confer "privileged status" on Israel in its relations with the European Union can become reality. In this context, the European People's Party calls for the intensification and expansion of the Barcelona process.

538 Europe and Immigration

Europe is a prosperous region enjoying democratic rights and fundamental freedoms and as such is an attractive target for immigration. It directly borders regions with a spiralling population growth, poverty and shortage of water, in which human rights are frequently violated. The migration problems will never be solved if Europe does not follow an intensified policy of development cooperation, including the establishment of a free trade area with the countries of the Mediterranean, and more generally, under the conditions laid down under the MEDA Programme , access to the European market for export products from developing countries, and of promotion of democracy, good governance and the rule of law.

The EPP calls for a Europe-wide law on immigration and the harmonisation of the way that we deal with newcomers to Europe. The EPP believes that we should help people to live and prosper in their own regions. This implies increased aid and investment in poorer areas and practical assistance to improve the basic standard of living in all such areas. The EPP believes that the transfer of knowledge and technology and investment in developing economies is very important in preventing massive migration flows which produce populations without roots and potential marginalization.

539. Common Foreign and Security Policy

540. European Foreign, Security and Defence Policy should be anchored within the single institutional framework of the EU. However, this policy area requires particular flexibility, above all in the military field. A common foreign policy must not fail on account of imprecise institutional provisions on when unanimity or majority voting should apply. A revision of the Amsterdam Treaty is therefore required. A Common European Foreign, Security, and Defence Policy does not signify the end of national foreign policy. The EU should adopt framework and strategic policies that must be

elaborated and defined through the community method. It is then for Member States to flesh out this framework and contribute their own particular skills to the process.

541. Member States wishing to act militarily must be able to do so in the EU's name provided that they obtain a quorum (yet to be specified, but different from the present one). All Members of the Union must act with solidarity, including financial solidarity.

600. VI New approaches and firm values

601. New questions will always arise in the European project, along with unexpected developments, new opportunities, and new horizons. Given the dynamism of the present times, concrete political choices may change. The EPP's firm values are more than ever essential, a clear beacon and frame of reference: to distinguish between what to conserve, what to improve, what to avoid, and what to combat.

602. We ultimately derive our strength and motivation from our values(freedom and responsibility, dignity of the human person, solidarity, justice and the rule of law), which are a whole vision of life, and cannot be separated from each other.

603. Freedom and responsibility of persons

604. We see men and women as responsible persons, each endowed with a unique and transcendent dignity. In order to fulfil themselves, people need freedom, autonomy and responsibility. Each person must be guaranteed the right to develop and use his or her gifts, talents, and abilities to create a meaningful life for themselves, their neighbours, and their environment. For us, this not only relates to their material, but to their cultural and spiritual needs as well. Each person has rights and obligations to receive and to give responsibility and care to and from other people, their communities and their societies, if necessary, be provided for by the government. This fundamental dignity gives each person the inalienable right to conduct his life in freedom. It also gives each person the duty to accept responsibility for his life and actions. This responsibility, which follows upon our free-

dom, must be shown towards all fellow human beings, to past and to future generations.

605. An essential part of this freedom and responsibility is the possibility for people to live and to organise themselves in civil society according to their fundamental beliefs and convictions. In the economy, the market is an instrument which enables people to deploy their gifts and talents on the basis of their own freedom and responsibility. The market should be free and responsible, both in a social sense and in its approach towards the natural environment. Civil society, the market, and the government are not ends in themselves, but ultimately serve to foster freedom, responsibility, and the dignity of the human person. The government should not take over those tasks that can best be performed by society itself. The subsidiarity principle must inform all decision-making in the European Union.

606. Fundamental Equality

607. All human beings, men and women, are equal and endowed with the same dignity. Relations between people should be based on respect for this equality, and on fundamental equality before the law. Thus, they are enabled to develop in freedom and responsibility according to their talents and abilities, irrespective of origin, age, race, nationality, sexual orientation, religion, conviction, social status or physical abilities. This starts at the very beginning of human life and continues until its fragile end.

608. Solidarity

609. As human beings we are by nature dependent on society and on one another. Every citizen has a duty to contribute to the building of society. The EPP will therefore seek to preserve and strengthen the sense of civic community and solidarity among all social groups and all individuals. On the basis of this view of society we will work to strengthen the smaller units of community life; from the individual and the family to the associations of civil society.

610. Solidarity, the care of the human person for his or her neighbour, is an indispensable part of the freedom and responsibility of the person as it is of government. Solidarity begins with the protection of those who are weakest, by civil society as well as by public institutions. Solidarity should colour

the relationships between people and their communities. Solidarity is no less relevant to relationships with people further away and with future generations. This includes respect for the integrity of Creation. We reject concepts that consider a person as an atomised individual guided by self-interest alone, as we also reject ideologies which think solidarity can only work through state institutions.

Europe must also deepen its common identity and solidarity in order to promote its social and economic cohesion and reduce the disparities between the levels of development of the various regions. The European People's Party is firmly persuaded that all the regions of the EU must have the same opportunities to participate in the development of the Union. This identity and solidarity must be extended to the candidate states.

611. Justice

612. Society can and should be shaped in freedom, responsibility, and solidarity. The value of justice particularly applies to the role of government, whether at the local, regional, national or at the European level. It is the basic task of government to create the conditions thanks to which persons, civil society, and the actors in the market, can optimally enjoy their freedom, responsibility, and solidarity on the basis of equality. This not only entails respect for civil and political rights and liberties, but the economic, social, and cultural preconditions for a life in dignity as well.

613. The law is the instrument government use to promote justice. Democratic politics - free elections, open and representative dialogue - is the best way through which a just order may come about. Genuine democracy can therefore never be incompatible with the rule of law. Accountability and integrity in political behaviour are essential safeguards. Democracy should be as close to the concerns of the people as possible. Only where the scale of problems so requires should a task be taken to a higher level. In many fields, European integration is an essential contribution to promote justice. Europe is not an end in itself, but is our contribution to a more peaceful and just global order.

A society of justice must provide for the safety of its citizens. Laws must exist and be upheld to combat criminality and corruption. But a good and safe society is first and foremost created through the upbringing of new generations which must be lead by certain principles and values. Parents and schools share the responsibility of teaching our children about what is right and wrong and what is the meaning of democracy, solidarity and equality. A "valueless" upbringing and education causes disorientation and distrust.

614. Subsidiarity

615. Society and the State have to serve the human person and the common good. Human persons and communities must have the right to accomplish what they can by their own initiative. What smaller scale and independent organizations cannot do has to be assigned to a greater and higher association (region, state, supranational organization). Therefore, subsidiarity is also the key principle of decentralisation, federalism, and European integration. Every social activity is by nature subsidiary. Politics must support the activities of human persons, families, and the intermediate society, not destroy or absorb them. In order to allow individuals and their free associations to develop to the maximum in the social field, the States - in all possible fields - should employ the resources saved for supporting a policy aimed at strong fiscal deduction for those citizens who pay contributions to socially characterized private bodies which work in the cultural, political, scientific and charitable fields.

616. Sustainability

617. Society must safeguard the opportunities of coming generations. Politics have to bear in mind sustainability not only of the environment, but also of public finance, pension schemes, and solidarity. We are as strongly aware of their heritage as of their future.

Chronology[308]

Early 1920s	Luigi Sturzo attempts to establish an 'International' of people's parties.
12–13 December 1925	The first meeting of Christian Democratic and people's parties' representatives is held in Paris.
1925–1926	The Secrétariat international des partis démocratiques d'inspiration chrétienne (SIPDIC) is founded, with the Secretariat in Paris.
October 1932	The last SIPDIC Congress is held in Cologne, chaired by Konrad Adenauer; a declaration is issued on a European common market.
1939	SIPDIC ceases to exist.
1940–1941	The International Christian Democratic Union (ICDU) is founded in London by Luigi Sturzo and British and exiled politicians.

---

[308] Membership status of parties is only mentioned once, with ordinary membership taking precedence over associate membership and associate membership taking presedence over observer membership. Re-election of presidents and secretary generals is not mentioned.

27 February– *2 March 1947*	A meeting of Western European party delegates is held in Lucerne (Switzerland); a decision is made to establish the Nouvelles équipes internationales (NEI).
23 April 1947	Organización Demócrata Cristiana de América (ODCA) is founded in Montevideo (Uruguay).
31 May– *3 June 1947*	The founding meeting of the NEI is held in Chaudfontaine (Belgium), with the Secretariat in Brussels (moved to Paris in 1950 and to Rome in 1964).
November 1947– *1955*	Secret meetings take place between leading French and German Christian Democrats in Geneva, known as the Geneva Circle.
7–11 May 1948	The NEI participates in The Hague Congress.
20 June 1950	The Christian Democratic Union of Central Europe (CDUCE) is founded in New York.
11 September *1952*	The first meeting of the Christian Democratic deputies in the Common Assembly of the European Coal and Steel Community (ECSC) takes place.
16 June 1953	The first official meeting of the Christian Democratic Group in the Common Assembly takes place.
July 1961	The Christian Democratic World Union (CDWU) is founded in Santiago (Chile).
September 1964	The name CDWU is changed to Christian Democrat International (CDI).
9–12 December *1965*	The NEI ceases to exist, and the European Union of Christian Democrats (EUCD) is founded, with Mariano Rumor elected President and Leo Tindemans elected Secretary General at a Congress in Taormina (Italy). The EUCD Secretariat is based in Rome.

April 1972	The Political Committee of Christian Democratic Parties from Member States of the European Communities is established.
7–9 November 1973	Kai-Uwe von Hassel succeeds Mariano Rumor as EUCD President at a Congress in Bonn.
April 1974	Arnaldo Forlani succeeds Leo Tindemans as EUCD Secretary General.
September 1975– February 1976	An ad hoc working group, chaired jointly by Hans August Lücker, Chair of the Christian Democratic Group in the European Parliament, and Wilfried Martens, President of the Flemish Christelijke Volkspartij (CVP), lays the groundwork for a European party.
29 April 1976	The Political Committee, meeting in Brussels, decides to found the European People's Party: Federation of Christian Democratic Parties of the European Community.
8 July 1976	The European People's Party (EPP) is founded by the Political Bureau of the EUCD at its meeting in Luxembourg; Belgian Prime Minister Leo Tindemans is elected President; founding members are the CVP and the Parti social chrétien (PSC) from Belgium, the Christlich Democratische Union (CDU) and the Christlich-Soziale Union (CSU) from Germany, the Centre des démocrates sociaux (CDS) from France, the Democrazia Cristiana (DC) from Italy, Fine Gael (FG) from Ireland, the Chrëschtlech-Sozial Volkspartei (CSV) from Luxembourg and the Christen Democratisch Appèl (CDA) from the Netherlands. The EPP Secretariat is based in Brussels.

6–7 March 1978	The first EPP Congress in Brussels adopts the programme 'Together towards a Europe for Free People', prepared by rapporteurs Hans August Lücker and Wilfried Martens. Jean Seitlinger of France is elected Secretary General by the EPP's Political Bureau.
14 March 1978	The name Christian Democratic Group is changed to Christian Democratic Group (Group of the European People's Party).
24 April 1978	The European Democratic Union (EDU) is founded in Salzburg (Austria). It has its Secretariat in Vienna.
6–7 June 1978	Giuseppe Petrilli succeeds Arnaldo Forlani as EUCD Secretary General at a Congress in Berlin.
22–23 February 1979	The second EPP Congress in Brussels adopts 'Striving Together for a Europe of Free Citizens', the election manifesto for the first direct elections to the European Parliament.
17 July 1979	The name Christian Democratic Group (Group of the European People's Party) is changed to Group of the European People's Party (Christian Democratic Group).
1–2 September 1980	The third EPP Congress in Cologne adopts the programme 'The Christian Democrats in the Eighties: Securing Freedom and Peace, Completing Europe'.
2–3 March 1981	Diogo Freitas do Amaral is elected EUCD President, Kai-Uwe von Hassel is elected Vice-President.
6–8 December 1982	The fourth EPP Congress in Paris adopts the programme 'Establishing Peace – Preserving Freedom – Uniting Europe'.
1983	Néa Dēmokratía (ND) from Greece becomes an EPP member.

November 1983	Giulio Andreotti succeeds Diogo Freitas do Amaral as EUCD President; Thomas Jansen becomes Secretary General of the EPP and the EUCD; the EUCD Secretariat moves to Brussels.
26 November 1983	The first EPP summit is held in Palais d'Egmont at the initiative of Helmut Kohl.
2–4 April 1984	The fifth EPP Congress in Rome adopts the Action Programme for the second parliamentary term of the European Parliament.
1985	Piet Bukman of the Netherlands succeeds Leo Tindemans as EPP President; Emilio Colombo of Italy becomes EUCD President.
10–12 April 1986	The sixth EPP Congress in The Hague adopts 'EPP's Tenth Anniversary: Striving for a People's Europe'.
1986	The Centro Democrático e Social (CDSp) from Portugal and the Partido Demócrata Popular (PDP), the Unió Democràtica de Catalunya (UDC) and the Partido Nacionalista Vasco (PNV) from Spain become EPP members.
31 March 1987	Jacques Santer succeeds Piet Bukman as EPP President.
7–8 November 1988	The seventh EPP Congress adopts 'On the People's Side' as the Action Programme for the third parliamentary term of the European Parliament.
1990	The sub-name is changed to 'Christian Democrats'.
10 May 1990	Wilfried Martens succeeds Jacques Santer as EPP President.
15–16 November 1990	The eighth EPP Congress in Dublin adopts 'For A Federal Constitution for the European Union.'

1991	The Partido Popular (PP) from Spain becomes an EPP member.
1 May 1992	The British and Danish Conservatives join the EPP Group.
June 1992	The Christian Democratic Union of Central Europe (CDUCE) is dissolved during a Congress in Warsaw (Poland); revised EUCD statutes come into force.
12–14 November 1992	The ninth EPP Congress in Athens adopts the Basic Programme.
1993	Det Konservative Folkeparti and Kristeligt Folkeparti (since 2003 the Kristendemokraterne) from Denmark become EPP members; the Italian Südtiroler Volkspartei (SVP) and the Partito Democratico Cristiano Sammarinese (PDCS) from San Marino become observers; the Centro Democrático e Social (CDSp) from Portugal is expelled.
January 1993	Wilfried Martens succeeds Emilio Colombo as EUCD President.
8–10 December 1993	The tenth EPP Congress in Brussels adopts 'Europe 2000: Unity in Diversity' as the Action Programme for the fourth parliamentary term of the European Parliament.
1994	Three Italian parties – Partito Popolare Italiano (PPI), Centro Cristiano Democratico (CCD) and Cristiani Democratici Uniti (CDUi) – succeed Democrazia Cristiana (DC) as EPP members; Klaus Welle succeeds Thomas Jansen as EPP and EUCD Secretary General.

February 1995	The Österreichische Volkspartei (ÖVP) from Austria, Moderata Samlingspartiet or Moderaterna (MS) and Kristdemokratiska Samhällspartiet (KDS) from Sweden, and Kansallinen Kokoomus (KOK) from Finland become EPP members; Høyre from Norway becomes an associate member.
6–7 November 1995	The eleventh EPP Congress in Madrid adopts the programme 'EPP – Force of the Union'.
1996	The Partido Social Democrata (PSD) from Portugal becomes an EPP member.
9–11 November 1997	The twelfth EPP Congress in Toulouse adopts the programme 'We Are All Part of One World'.
1998	Rinnovamento Italiano (RI) from Italy becomes an EPP member.
4–6 February 1999	The thirteenth EPP Congress in Brussels adopts 'On the Way to the 21st Century', the Action Programme for the fifth parliamentary term of the European Parliament; Alejandro Agag succeeds Klaus Welle as EPP Secretary General; the EUCD and EPP officially merge.
July 1999	The EPP Group changes its name to Group of the European People's Party (Christian Democrats) and European Democrats (EPP-ED Group).
December 1999	Italy's Forza Italia (FI) becomes an EPP member.
2000	EDU Secretariat moves to Brussels.
2001	France's Rassemblement pour la République (RPR), a year later succeeded by Union pour un mouvement populaire (UMP), becomes an EPP member.
11–13 January 2001	The fourteenth EPP Congress in Berlin adopts a new basic document, 'A Union of Values'.

2002	The Italian Unione di Centro (UdC) succeeds Cristiani Democratici Uniti (CDUi) and Centro Cristiano Democratici (CCD) as an EPP member; Demokratski centar (DCc) from Croatia becomes an observer.
17–18 October 2002	The fifteenth EPP Congress in Estoril (Portugal) adopts 'A Constitution for a Strong Europe'; the EDU ceases to exist; Antonio López-Istúriz succeeds Alejandro Agag as EPP Secretary General.
2003	The Albanian Partia Demokratike e Shqipërisë (PDa) becomes an observer; the Serbian G17 PLUS and Demokratska stranka Srbije (DSS) become associate members.
2004	Stranka demokratske akcije (SDA), Hrvatska demokratska zajednica Bosne i Hercegovine (HDZBiH) and Partija demokratskog progresa (PDPbih) from Bosnia and Herzegovina become observers; Croatia's Hrvatska demokratska zajednica (HDZ) becomes an associate member.
4–5 February 2004	The sixteenth EPP Congress in Brussels adopts 'The EPP: Your Majority in Europe', the Action Programme for the sixth parliamentary term of the European Parliament.
May 2004	Dimokratikos Synagermos (DISY) from Cyprus, Křesťanská a demokratická unie–Československá strana lidová (KDU-ČSL) from the Czech Republic, Erakond Isamaaliit–Pro Patria Union and Erakond Res Publica from Estonia, Fidesz and Magyar Demokrata Fórum (MDF) from Hungary, Jaunais Laiks (JL) and Tautas Partija (TP) from Latvia, Tėvynės Sąjunga (TS) and Lietuvos krikščionys demokratai (LKD) from Lithuania, Partit Nazzjonalista (PN) from Malta, Platforma Obywatelska (PO) and Polskie Stronnictwo Ludowe (PSL)

from Poland, Kresťanskodemokratické hnutie (KDH), Slovenská demokratická a Kresťanská únia–Demokratická strana (SDKÚ-DS) and Strana maďarskej koalície–Magyar Koalíció Pártja (SMK-MKP) from Slovakia and Slovenska demokratska stranka (SDS), Nova Slovenija–Krščanska ljudska stranka (NSi-KLS) and Slovenska ljudska stranka (SLS) from Slovenia become EPP members.

July 2004	The Italian PPI and the French UDF leave the EPP and its Group in the European Parliament.
2005	Adalet ve Kalkınma Partisi (AKP) from Turkey, Partidul Popular Creştin Democrat (PPCD) from Moldova, and Narodnyi Soyuz Nasha Ukrayina (NSNU) and Narodnyi rukh Ukraïny (RUKH) from Ukraine become EPP observers.
2006	Bielaruski narodny front (BPF) and Abjadnanaja hramadzianskaja partyja Bielarusi (UCP) from Belarus become observers.
30–31 March 2006	The seventeenth EPP Congress in Rome adopts 'For a Europe of the Citizens: Priorities for a Better Future'.
2007	The Bulgarian Săjuz na Demokratičnite Sili (UDFb), Zemedelski Naroden Sajuz (ZNS), Demokratičeska Partija (DP) and Demokrati za Silna Bălgarija (DSB); the Romanian Partidul Democrat-Liberal (PD-L), Uniunea Democrată Maghiară din Romaniai–România Magyar Demokrata Szövetség (RMDSZ) and Partidul Naţional Ţărănesc Creştin Democrat (PNŢCD); and the Hungarian Kereszténydemokrata Néppárt (KDNP) become EPP members; Hrvatska seljačka stranka (HSS) from Croatia becomes an associate member; the Hungarian Savez vojvođanskih Mađara–Vajdasági Magyar Szövetség (VMSZ) becomes an observer member.

13 September 2007	The Centre for European Studies (CES) is founded in Brussels.
2008	Graždani za Evropeĭsko Razvitie na Bălgarija (GERB) from Bulgaria becomes an EPP member; Batkivshchyna from Ukraine and Ertiani Natsionaluri Modzraoba (UNM) from Georgia become EPP observers.
29–30 March 2009	The eighteenth EPP Congress in Warsaw adopts 'Strong for the People', the Action Programme for the seventh parliamentary term of the European Parliament; the Centro Democrático e Social–Partido Popular (CDS-PP) from Portugal becomes an EPP member.
June 2009	Il Popolo della Libertà (Italy) replaces Forza Italia as an EPP member.
July 2009	The British Conservatives and the Czech Občanská demokratická strana (ODS) leave the EPP-ED Group; the name changes to EPP Group.
9–10 December 2009	The nineteenth EPP Congress in Bonn adopts the programme 'The Social Market Economy in a Globalised World', as well as a resolution on climate change.
May 2010	The Latvian Pilsoniskā Savienība (PS) becomes an EPP member.
September 2010	The Vnatrešna Makdonska Revolucionerne Organizacija–Demokratska Partija za Makedonsko Nacionalno Edinstvo (VMRO-DPMNE) from FYROM becomes an associate member.
February 2011	The Czech Tradice, Odpovědnost, Prosperita 09 (TOP09) becomes a full member and the Moldovan Partidul Liberal Democrat din Moldova (PLDM) becomes an observer member.

References

Ashford S, Timms N (1992). *What Europe Thinks: A Study of Western European Values*. Aldershot.

Bacharan-Gressel N (1993). 'Les organisations et les associations pro-européennes'. In: Berstein S, Mayeur J-M, Milza P, *Le MRP et la construction européene*. Brussels, pp. 41ff.

Barbi P (1985). *Napoli–Strasburgo e ritorno. I cinque anni al Parlamento Europeo*. Napoli.

Becker W, Morsey R (eds.) (1988). *Christliche Demokratie in Europa: Grundlagen und Entwicklungen seit dem 19. Jahrhundert*. Cologne and Vienna.

Bichet R (1980). *La Démocratie chrétienne en France: Le Mouvement républicain populaire*. Besançon.

Böx H (1988). 'Demokratie im christlichen Europa'. In: Jenninger P et al. (eds.), *Unverdrossen für Europa: Festschrift für Kai-Uwe von Hassel*. Baden-Baden, p. 144.

Buchstab G (ed.) (1986). *Protocols of the CDU Federal Leadership 1950–1953*. Stuttgart.

Bukman P et al. (1988). *Tradition und Aktualität der Bemühungen um eine 'Doktrin'*. Vol. 1, EPP series Geistige und historische Grundlagen Christlichdemokratischer Politik. Melle.

Chenaux P (1990). *Une Europe vaticane? Entre le plan Marshall et les traités de Rome*. Brussels.

Chenaux P (1992). 'Les Nouvelles équipes internationales'. In: Pistone S (ed.), *I Movimenti per l'unità Europea dal 1945 al 1954: Atti del convegno internazionale Pavia 19–20–21 ottobre 1989.* Milano: Fondazione Europea Luciano Bolis, pp. 237–52.

Chenaux P (1995). 'La contribution belge à la démocratie chrétienne internationale' (unpublished paper).

Colombo E (1990). 'Internationale Präsenz der Christlichen Demokraten'. In: Fraktion der Europäischen Volkspartei, *Zur Geschichte der christlich-demokratischen Bewegung in Europa.* Melle, p. 80.

Delbreil J-C (1993). 'Le MRP et la construction européenne: résultats, interprétations, et conclusions d'une enquête écrite et orale'. In: Berstein S, Mayeur J-M, Milza P, *Le MRP et la construction européene.* Brussels, pp. 309–63.

Delbreil J-C (1993). 'Les Démocrates d'inspiration chrétienne et les problèmes européens dans l'entre-deux-guerres'. In: Berstein S, Mayeur J-M, Milza P, *Le MRP et la construction européene.* Brussels, pp. 5–39.

Dörpinghaus B (1976). 'Die Genfer Sitzungen – Erste Zusammenkünfte führender christlich-demokratischen Politiker im Nachkriegseuropa'. In: Blumenwitz D et al. (eds.), *Konrad Adenauer und seine Zeit: Politik und Persönlichkeit des ersten Bundeskanzlers.* Vol. 1: *Beiträge von Weg- und Zeitgenossen.* Stuttgart, pp. 358–65.

Durand J-D (1993). 'Les rapports entre le MPR et la Démocratie chrétienne italienne (1945–1955)'. In: Berstein S, Mayeur J-M, Milza P, *Le MRP et la construction européene.* Brussels, pp. 251ff.

Fogarty MP (1957). *Christian Democracy in Western Europe 1820–1953.* London.

Fontaine P (2009). *Voyage to the Heart of Europe 1953–2009. A History of the Christian-Democratic Group and the Group of the European People's Party in the European Parliament.* Brussels.

Fraktion der Europäischen Volkspartei (1985). *Liber Amicorum: Erinnerungen an Hans August Lücker zum 70. Geburtstag.* Bonn.

Fraktion der Europäischen Volkspartei (1990). *Zur Geschichte der christlich-demokratischen Bewegung in Europa.* Melle.

Gebauer B (1978). *Die europäischen Parteien der Mitte: Analysen und Dokumente zur Programmatik christlich-demokratischer und konservativer Parteien Westeuropas*, Handbücher der Politischen Akademie Eichholz, vol. 6. Bonn.

Gehler M (2001). 'Begegnungsort des Kalten Krieges. Der "Genfer Kreis" und die geheimen Absprachen westeuropäischer Christdemokraten (1947–1955)'. In: Gehler M, Kaiser W, Worhnout H (eds.), *Christdemokratie in Europa im 20. Jahrhundert*. Vienna, pp. 642–94.

Gisch H (1990). 'Die europäischen Christdemokraten (NEI)'. In: Loth W (ed.), *Die Anfänge der europäischen Integration 1945–1950*. Bonn, pp. 227–36.

Grabitz E, Läufer T (1980). *Das Europäische Parlament*. Bonn.

Hahn KJ, Fugmann F (1976). 'Die Europäische Christlich-Demokratische Union zwischen europäischen Anspruch und nationalen Realitäten'. In: Wessels W, *Zusammenarbeit der Parteien in Westeuropa. Auf dem Weg zu einer neuen politischen Infrastruktur?* Bonn, pp. 255ff.

Hahn KJ (ed.), *Zusammenarbeit der Parteien in Westeuropa: Auf dem Weg zu einer neuen politischen Infrastruktur?* Bonn, pp. 304–31.

Hallstein W (1969). *Der Unvollendete Bundesstaat*. Düsseldorf.

Hanschmidt A (1988). 'Eine christlich-demokratische "Internationale" zwischen den Weltkriegen: Das "Secrétariat International des Partis Démocratiques d'Inspiration Chrétienne" in Paris'. In: Becker W, Morsey R (eds.), *Christliche Demokratie,* pp. 158–88.

Horner F (1981). *Konservative und christ-demokratische Parteien in Europa: Geschichte, Programmatik, Strukturen*. Vienna and Munich.

Hrbek R (1981). 'Die europäischen Parteienzusammenschlüsse'. In: Weidenfeld W, Wessels W (eds.), *Jahrbuch der Europäischen Integration 1980*. Bonn, p. 261.

Jansen T (1991). 'Europäische Christdemokraten überprüfen ihre Doktrin'. *Politische Meinung* 256/36, March 1991, pp. 66–72.

Jansen T (1992). 'Zur Entwicklung Supranationaler Europäischer Parteien'. In: Gabriel OW et al. (eds.), *Der Demokratische Verfassungsstaat*.

Theorie, Geschichte, Probleme. Festschrift für Hans Buchheim. Munich, pp. 241ff.

Jost S (1994). *Die Politische Mitte Spaniens: Von der Union de Centro zum Partido Popular*, Saarbrücker Politikwissenschaft, vol. 18. Frankfurt am Main.

Karnofski E-R (1982). *Parteienbünde vor der Europa-Wahl 1979: Integration durch gemeinsame Wahlaussagen?* Bonn.

Khol A (1978). 'Europäische Demokratische Union (EDU): Die europäischen Parteiengruppe der fortschrittlichen Mitte'. *Österreichische Monatshefte: Zeitschrift für Politik* 5 (1978), pp. 4ff.

Khol A, Wintoniak A (2000). 'Die Europäische Demokratische Union (EDU)'. In: Veen H-J (ed.), *Christlich-Demokratische und Konservative Parteien in Westeuropa 5,* Studien zu Politik, vol. 31. Paderborn: Konrad Adenauer Foundation, pp. 403–57.

Lamberts E (2002). *The Black International 1870–1878: The Holy See and Militant Catholicism in Europe.* Leuven.

Letamendia P, 'L'Union européenne démocrate chrétienne'. In: Portelli H, Jansen T (eds.), *La Démocratie chrétienne: force internationale.* Nanterre, pp. 55–63.

Lücker HA, Hahn KJ (1987). *Christliche Demokraten bauen Europa.* Bonn.

Martens W (1994). *L'une et l'autre Europe, Discours européen 1990–94.* Brussels.

Martens W (2009). *Europe: I Struggle, I Overcome.* Brussels.

Mayeur J-M (1980). *Des Partis catholiques à la démocratie chrétienne.* Paris.

Metsola UE (2000). *Towards the Majority: An Analysis of the Rapprochement between the European People's Party and the European Democrat Union.* Helsinki.

Möller H (2007). 'Hanns Seidels christliches Menschenbild'. In: Zehetmair H (Ed.), *Politik aus christlicher Verantwortung.* Wiesbaden.

Nassmacher K-H (1972). *Demokratisierung der Europäischen Gemeinschaften*. Bonn.

Niedermayer O (1983). *Europäische Parteien? Zur grenzüberschreitenden Interaktion politischer Parteien im Rahmen der Europäischen Gemeinschaft*. Frankfurt am Main/New York.

Papini R (1986). 'Les débuts des Nouvelles équipes internationales'. In: Portelli H, Jansen T (eds.), *La Démocratie chrétienne: force internationale*. Nanterre, pp. 31–40.

Papini R (1988). *L'Internationale démocrate-chrétienne: la coopération entre les partis démocrates-chrétiens de 1925 à 1986*. Paris.

Papini R (1995). 'Il corraggio della democrazia: Luigi Sturzo e l'Internazionale Popolare tra le due guerre' (unpublished paper).

Papini R (1997). *The Christian Democrat International*. Lanham.

Rumor M (1990). 'Die gemeinsame Aktion der Christlichen Demokraten in Europa'. In: Fraktion der Europäischen Volkspartei, *Zur Geschichte der christlich-demokratischen Bewegung in Europa*. Melle, p. 89.

Schreiner R (1993). 'La politique europèenne de la CDU relative à la France et au MPR des anneés 1945–1966'. In: Berstein S, Mayeur J-M, Milza P, *Le MRP et la construction européene*. Brussels, pp. 275ff.

Siniewicz K (1986). 'L'activité internationale des démocrates-chrétiennes de l'Europe centrale'. In: Portelli H, Jansen T (eds.), *La Démocratie chrétienne: force internationale*. Nanterre, pp. 233ff.

Stammen T (1977). *Parteien in Europa. Nationale Parteiensysteme. Transnationale Parteienbeziehungen. Konturen eines europäischen Parteiensystems*. Munich.

ten Napel HM (1997). 'Van het continentale naar het angelsaksiche model van christen-democratie? Over de problematische europeanisering van de christen-democratische politiek'. In: *Jaarboek Documentatiecentrum Nederlandse Politieke Partijen*. Groningen, p. 233.

Van Hecke S (2004). 'A Decade of Seized Opportunities: Christian Democracy in the European Union'. In Van Hecke S, Gerard E (eds.), *Christian Democratic Parties in Europe since the End of the Cold War*. Leuven, pp. 269–95.

Van Hecke S (2006). 'On the Road towards Transnational Parties in Europe: Why and How the European People's Party Was Founded'. *European View* 3 (Spring 2006).

Van Hecke S (2009). 'Europeanization and Political Parties: The *Partido Popular* and its Transnational Relations with the European People's Party'. *International Journal of Iberian Studies* 22, pp. 109–124.

von der Bank E, Szabó K (eds.) (2006). *Robert Schuman Institute. Fifteen Years for Developing Democracy in Central and Eastern Europe*. Budapest.

von Petersdorff E (ed.) (1971). *Synopse der Parteiprogramme der Christlich-Demokratischen Parteien Westeuropas*. Wesseling.

Welle K (2000). 'Reform of the European People's Party 1995–1999'. In: Veen H-J (ed.), *Christlich-demokratische und konservative Parteien in Westeuropa 5*. Paderborn-Munich, Vienna, Zurich, 2000, pp. 541–66.

Wessels W (1976). *Zusammenarbeit der Parteien in Westeuropa. Auf dem Wege zu einer neuen politischen Infrastruktur?* Bonn.

Index

Printing: Ten Brink, Meppel, The Netherlands
Binding: Stürtz, Würzburg, Germany